STUDENT SOLUTIONS MANUAL
BY EDW. S. GINSBERG FOR
PHYSICS FOR SCIENTISTS AND ENGINEERS

VOLUME II

RICHARD WOLFSON | JAY M. PASACHOFF

Taken from:

Student Solutions Manual by Edw. S. Ginsberg for
Physics for Scientists and Engineers, Third Edition
by Richard Wolfson, Jay M. Pasachoff

Cover Art *Electr1* by Barry Cronin
Taken from:

Student Solutions Manual by Edw. S. Ginsberg
for
Physics, For Scientists and Engineers, Third Edition
by Richard Wolfson, Jay M. Pasachoff

A Pearson Education Company
Boston, Massachusetts 02116

This special edition published in cooperation with Pearson Custom Publishing.

Printed in the United States of America

10 9 8 7 6 5 4 3 2 1

ISBN 0-536-17010-X

2005460101

MT

Please visit our web site at *www.pearsoncustom.com*

PEARSON CUSTOM PUBLISHING
75 Arlington Street, Suite 300, Boston, MA 02116
A Pearson Education Company

CONTENTS

PREFACE

The decision to compile a manual of worked-out solutions to the end-of-chapter problems in Wolfson and Pasachoff's textbook arose from a consideration of the crucial importance of homework problems in the learning of physics. The process of applying physical concepts to construct explanations and answer questions is an excellent means of increasing one's understanding and confirming one's knowledge of those very concepts. (Learning physics without doing problems is like learning to swim without going in the water!) These solutions are designed to aid this process by supplementing, reinforcing, and sometimes replacing classroom discussions.

This *Student Solutions Manual* contains solutions to every odd-numbered problem in the text. In order for the reader to achieve the most benefit from using this manual, most of the solutions have been written in a form that requires his or her active participation. Such an actively engaged reader simultaneously reads and works through each solution with pencil and paper, calculator, and open textbook. Intermediate steps, such as algebraic manipulations, substitution of variables or data, diagrams, conversion of units, etc., are frequently and intentionally omitted. They must be supplied by an engaged reader, using the manual's solutions as guideposts and textual materials as references. Wolfson and Pasachoff intended their problems to help readers understand and use concepts presented in their textbook, and develop problem-solving skills. I have integrated references to the text into my solutions in fulfillment of this goal.

My solutions are not necessarily the only ones possible (or even the best, simplest, etc.). In some cases, alternate methods are indicated; in others, a different point of view from the text's is deliberately presented. Students should remember that the logical equivalence of different approaches to a problem has to be demonstrated by means other than just getting the same answer.

Many authors have enumerated helpful suggestions for students to follow when solving problems. My own version of such a list is the following:

- Carefully READ and understand the question. It may be necessary to visualize a situation or device, as described in the text, or as known from personal experience. The context of the question, i.e., the chapter or section in which it occurs, is often of help.
- THINK about how the quantities that determine the sought-for answer are physically related to the quantities that are either given in the question, or obtainable from them and other sources of data. Construction of a simple, physical, conceptual model, to represent the situation in the problem, may often be necessary.
- Write down physical equations involving the relevant quantities and relations, perhaps using approximations where appropriate, and SOLVE for the desired variables. For some problems, this is the most difficult step.
- CONSIDER the reasonableness of your answer. Does it make sense? Is it consistent with your initial expectations? Does it change suitably when the conditions in the problem are altered? If not, the previous steps may require repetition or verification.

In the real world of experimental science, numerical results reflect the precision of actual measurements and theoretical uncertainties in their interpretation. Such subjects are probably more appropriate to laboratory or advanced courses. By default, I have adopted the once-standard convention of regarding most numbers in problems as accurate to three significant figures, unless otherwise indicated by the context. I think common sense is a better guide than consistency at this level. On a different aspect of accuracy, I must admit to the responsibility for any errors which inevitably are present in a work of this size, and will make any corrections brought to my attention.

Although I personally have solved every problem in this manual, I acknowledge a great debt to other authors, my colleagues, former teachers, and students. Wolfson and Pasachoff have produced a scholarly, interesting, and well-written textbook with problems to match. I have tried to emulate their high standards in this accompanying manual of solutions, which I hope will be a practical aid to both students and instructors.

Edw. S. Ginsberg
University of Massachusetts Boston
Boston, MA 02125-3393
edw.ginsberg@umb.edu

Additional Supplements to Wolfson and Pasachoff,
Physics for Scientists and Engineers, Third Edition

Student Study Guide Volume 1, with *ActivPhysics1* (ISBN: 0-321-05148-3)
Student Study Guide Volume 2, with *ActivPhysics2* (ISBN: 0-321-05147-5)

PART 4 ELECTROMAGNETISM

CHAPTER 23 ELECTRIC CHARGE, FORCE, AND FIELD

ActivPhysics can help with these problems: Activities 11.1–11.8

Section 23-2: Electric Charge

Problem

1. Suppose the electron and proton charges differed by one part in one billion. Estimate the net charge you would carry.

Solution

Nearly all of the mass of an atom is in its nucleus, and about one half of the nuclear mass of the light elements in living matter (H, O, N, and C) is protons. Thus, the number of protons in a 65 kg average-sized person is approximately $\frac{1}{2}(65\text{ kg})/(1.67\times10^{-27}\text{ kg}) \approx 2\times10^{28}$, which is also the number of electrons, since an average person is electrically neutral. If there were a charge imbalance of $|q_{\text{proton}} - q_{\text{electron}}| = 10^{-9}e$, a person's net charge would be about $\pm2\times10^{28}\times10^{-9}\times 1.6\times10^{-19}\text{ C} = \pm3.2\text{ C}$, or several coulombs (huge by ordinary standards).

Problem

3. Protons and neutrons are made from combinations of the two most common quarks, the u quark and the d quark. The u quark's charge is $+\frac{2}{3}e$ while the d quark carries $-\frac{1}{3}e$. How could three of these quarks combine to make (a) a proton and (b) a neutron?

Solution

(a) The proton's charge is $1e = \frac{2}{3}e + \frac{2}{3}e - \frac{1}{3}e$, corresponding to a combination of *uud* quarks; (b) for neutrons, $0 = \frac{2}{3} - \frac{1}{3} - \frac{1}{3}$ corresponds to *udd*. (See Chapter 39, or Chapter 45 in the extended version of the text.)

Section 23-3: Coulomb's Law

Problem

5. If the charge imbalance of Problem 1 existed, what would be the approximate force between you and another person 10 m away? Treat the people as point charges, and compare the answer with your weight.

Solution

The magnitude of the Coulomb force between two point charges of 3.2 C (see solution to Problem 1), at a distance of 10 m, is $kq^2/r^2 = (9\times10^9\text{ N·m}^2/\text{C}^2)\times (3.2\text{ C}/10\text{ m})^2 = 9.22\times10^8\text{ N}$. This is approximately 1.45 million times the weight of an average-sized 65 kg person.

Problem

7. The electron and proton in a hydrogen atom are 52.9 pm apart. What is the magnitude of the electric force between them?

Solution

$a_0 = 52.9$ pm is called the Bohr radius. For a proton and electron separated by a Bohr radius, $F_{\text{Coulomb}} = ke^2/a_0^2 \simeq (9\times10^9\text{ N·m}^2/\text{C}^2)(1.6\times10^{-19}\text{ C}\div 5.29\times10^{-11}\text{ m})^2 = 8.23\times10^{-8}\text{ N}$.

Problem

9. Two charges, one twice as large as the other, are located 15 cm apart and experience a repulsive force of 95 N. What is the magnitude of the larger charge?

Solution

The product of the charges is $q_1q_2 = r^2F_{\text{Coulomb}}/k = (0.15\text{ m})^2(95\text{ N})/(9\times10^9\text{ N·m}^2/\text{C}^2) = 2.38\times10^{-10}\text{ C}^2$. If one charge is twice the other, $q_1 = 2q_2$, then $\frac{1}{2}q_1^2 - 2.38\times10^{-10}$ C and $q_1 = \pm21.8\ \mu$C.

Problem

11. A proton is on the x-axis at $x = 1.6$ nm. An electron is on the y-axis at $y = 0.85$ nm. Find the net force the two exert on a helium nucleus (charge $+2e$) at the origin.

Solution

A unit vector from the proton's position to the origin is $-\hat{\imath}$, so the Coulomb force of the proton on the helium nucleus is $\mathbf{F}_{p,He} = k(e)(2e)(-\hat{\imath})/(1.6 \text{ nm})^2 = -0.180\hat{\imath}$ nN. (Use Equation 23-1, with q_1 for the proton, q_2 for the helium nucleus, and the approximate values of k and e given.) A unit vector from the electron's position to the origin is $-\hat{\jmath}$, so its force on the helium nucleus is $\mathbf{F}_{e,He} = k(-e)(2e)(-\hat{\jmath})\div(0.85 \text{ nm})^2 = 0.638\hat{\jmath}$ nN. The net Coulomb force on the helium nucleus is the sum of these. (The vector form of Coulomb's law and superposition, as explained in the solution to Problems 15 and 19, provides a more general approach.)

Problem

13. A charge q is at the point $x = 1$ m, $y = 0$. Write expressions for the unit vectors you would use in Coulomb's law if you were finding the force that q exerts on other charges located at (a) $x = 1$ m, $y = 1$ m; (b) the origin; (c) $x = 2$ m, $y = 3$ m. Note that you don't know the sign of q. Why doesn't this matter?

Solution

A unit vector from $\mathbf{r}_q = (1 \text{ m}, 0)$, the position of charge q, to any other point $\mathbf{r} = (x, y)$ is $\hat{\mathbf{n}} = (\mathbf{r}-\mathbf{r}_q)\div|\mathbf{r}-\mathbf{r}_q| = (x - 1 \text{ m}, y)/\sqrt{(x-1 \text{ m})^2 + y^2}$. The sign of q doesn't affect this unit vector, but the signs of both charges do determine whether the force exerted by q is repulsive or attractive, i.e., in the direction of $+\hat{\mathbf{n}}$ or $-\hat{\mathbf{n}}$. (a) When the other charge is at position $\mathbf{r} = (1 \text{ m}, 1 \text{ m})$, $\hat{\mathbf{n}} = (0, 1 \text{ m})/\sqrt{0 + (1 \text{ m})^2} = (0, 1) = \hat{\jmath}$. (b) When $\mathbf{r} = (0, 0)$, $\hat{\mathbf{n}} = (-1 \text{ m}, 0)/\sqrt{(-1 \text{ m})^2 + 0} = (-1, 0) = -\hat{\imath}$. (c) Finally, when $\mathbf{r} = (2 \text{ m}, 3 \text{ m})$, $\hat{\mathbf{n}} = (1 \text{ m}, 3 \text{ m})/\sqrt{(1 \text{ m})^2 + (3 \text{ m})^2} = (1, 3)/\sqrt{10} = 0.316\hat{\imath} + 0.949\hat{\jmath}$.

Problem

15. A 9.5-μC charge is at $x = 16$ cm, $y = 5.0$ cm, and a -3.2-μC charge is at $x = 4.4$ cm, $y = 11$ cm. Find the force on the negative charge.

Solution

Denote the positions of the charges by $\mathbf{r}_1 = (16\hat{\imath} + 5\hat{\jmath})$ cm for $q_1 = 9.5\ \mu$C, and $\mathbf{r}_2 = (4.4\hat{\imath} + 11\hat{\jmath})$ cm for $q_2 = -3.2\ \mu$C. The vector from q_1 to q_2 is $\mathbf{r} = \mathbf{r}_2 - \mathbf{r}_1$, and a unit vector in this direction is $\hat{\mathbf{r}} = (\mathbf{r}_2 - \mathbf{r}_1)\div|\mathbf{r}_2 - \mathbf{r}_1|$. The vector form of Coulomb's law for the electric force of q_1 on q_2 is $\mathbf{F}_{12} = kq_1q_2(\mathbf{r}_2 - \mathbf{r}_1$ $|\mathbf{r}_2 - \mathbf{r}_1|^3$. (This gives the Coulomb force betw point charges, as a function of their positions, convenient form to memorize because of its dir applicability.) Substituting the given values for problem, we find:

$$\mathbf{F}_{12} = \left(\frac{9\times10^9 \text{ N}\cdot\text{m}^2}{\text{C}^2}\right)(9.5\ \mu\text{C})(-3.2\ \mu\text{C}$$
$$\times\frac{(4.4\hat{\imath} + 11\hat{\jmath} - 16\hat{\imath} - 5\hat{\jmath}) \text{ cm}}{[(4.4-16)^2 + (11-5)^2]^{3/2} \text{ cm}^3}$$
$$= (14.2\hat{\imath} - 7.37\hat{\jmath}) \text{ N},$$

with magnitude 16.0 N and direction $\theta = -27.$ the x-axis (negative angle measured CW).

Problem

17. A 60-μC charge is at the origin, and a seco charge is on the positive x-axis at $x = 75$ c third charge placed at $x = 50$ cm experienc net force, what is the second charge?

Solution

In order for the net force to be zero at a positi between the first two charges, they must both same sign, i.e., $q_1 = 60\ \mu$C at $x_1 = 0$ and $q_2 >$ $x_2 = 75$ cm. (Then the separate forces of the fi charges on the third are in opposite directions. Therefore, for the third charge q_3 at $x_3 = 50$ cr $F_{3x} = kq_3[q_1(x_3 - x_1)^{-2} - q_2(x_2 - x_3)^{-2}] = kq$ $[60\ \mu\text{C}(50 \text{ cm})^{-2} - q_2(25 \text{ cm})^{-2}] = 0$ implies q_2 $60\ \mu\text{C}(25/50)^2 = 15\ \mu$C.

Problem

19. In Fig. 23-39 take $q_1 = 68\ \mu$C, $q_2 = -34\ \mu$ $q_3 = 15\ \mu$C. Find the electric force on q_3.

Solution

Denote the positions of the charges by $\mathbf{r}_1 = \hat{\jmath}$, and $\mathbf{r}_3 = 2\hat{\imath} + 2\hat{\jmath}$ (distances in meters). The vec form of Coulomb's law (in the solution to Probl

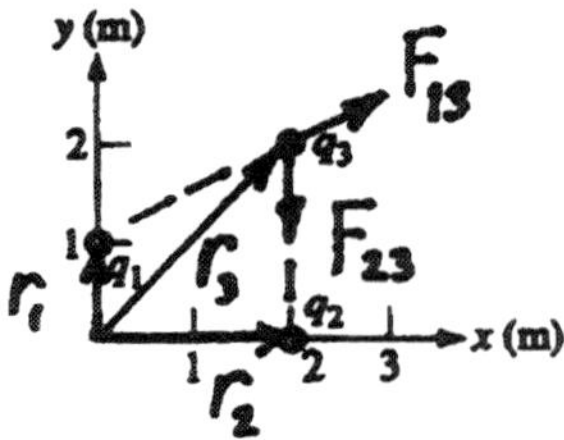

FIGURE 23-39 Problem 19 Solution.

and the superposition principle give the net electric force on q_3 as:

$$\mathbf{F}_3 = \mathbf{F}_{13} + \mathbf{F}_{23} = \frac{kq_1q_3(\mathbf{r}_3 - \mathbf{r}_1)}{|\mathbf{r}_3 - \mathbf{r}_1|^3} + \frac{kq_2q_3(\mathbf{r}_3 - \mathbf{r}_2)}{|\mathbf{r}_3 - \mathbf{r}_2|^3}$$
$$= (9\times10^9 \text{ N})(15\times10^{-6})[(68\times10^{-6})(2\hat{\imath}+\hat{\jmath})/5\sqrt{5} + (-34\times10^{-6})2\hat{\jmath}/8]$$
$$= (1.64\hat{\imath} - 0.326\hat{\jmath}) \text{ N},$$

or $F_3 = \sqrt{F_{3x}^2 + F_{3y}^2} = 1.67$ N at an angle of $\theta = \tan^{-1}(F_{3y}/F_{3x}) = -11.2°$ to the x-axis.

Problem

21. Four identical charges q form a square of side a. Find the magnitude of the electric force on any of the charges.

Solution

By symmetry, the magnitude of the force on any charge is the same. Let's find this for the charge at the lower left corner, which we take as the origin, as shown. Then $\mathbf{r}_1 = 0$, $\mathbf{r}_2 = a\hat{\jmath}$, $\mathbf{r}_3 = a(\hat{\imath}+\hat{\jmath})$, $\mathbf{r}_4 = a\hat{\imath}$, and

$$\mathbf{F}_1 = kq^2\left[\frac{\mathbf{r}_1 - \mathbf{r}_2}{|\mathbf{r}_1 - \mathbf{r}_2|^3} + \frac{\mathbf{r}_1 - \mathbf{r}_3}{|\mathbf{r}_1 - \mathbf{r}_3|^3} + \frac{\mathbf{r}_1 - \mathbf{r}_4}{|\mathbf{r}_1 - \mathbf{r}_4|^3}\right]$$
$$= kq^2\left[\frac{-a\hat{\jmath}}{a^3} - \frac{a(\hat{\imath}+\hat{\jmath})}{2\sqrt{2}a^3} - \frac{a\hat{\imath}}{a^3}\right]$$
$$= -\frac{kq^2}{a^2}(\hat{\imath}+\hat{\jmath})\left(1 + \frac{1}{2\sqrt{2}}\right),$$

(Use the vector form of Coulomb's law in the solution to Problem 15, and the superposition principle.) Since $|\hat{\imath}+\hat{\jmath}| = \sqrt{2}$, $|\mathbf{F}_1| = (kq^2/a^2)\sqrt{2}(1 + 1/2\sqrt{2}) = (kq^2/a^2)(\sqrt{2} + \frac{1}{2}) = 1.91kq^2/a^2$.

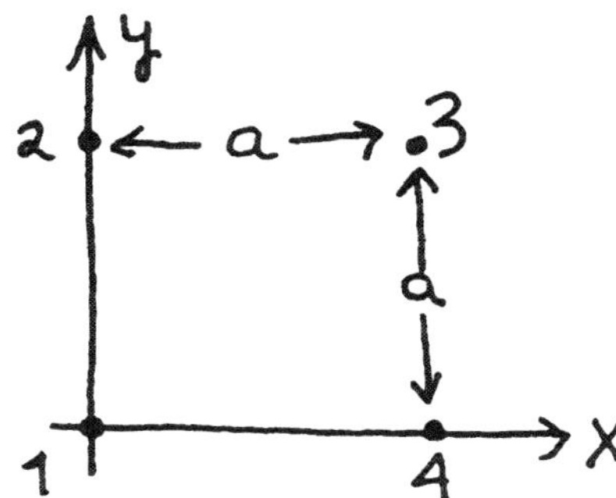

Problem 21 Solution.

Problem

23. Three charges lie in the x-y plane: $q_1 = 55$ μC at $x = 0$, $y = 2.0$ m; q_2 at $x = 3.0$ m, $y = 0$; and q_3 at $x = 4.0$ m, $y = 3.0$ m. If the force on q_3 is $8.0\hat{\imath} + 15\hat{\jmath}$ N, find q_2 and q_3.

Solution

Using the vector form of Coulomb's law explained in the solution to Problem 15, and superposition, we can write the force on q_3 as $\mathbf{F}_3 = kq_3[q_1(\mathbf{r}_3 - \mathbf{r}_1)\times |r_3 - r_1|^{-3} + q_2(\mathbf{r}_3 - \mathbf{r}_2)|\mathbf{r}_3 - \mathbf{r}_2|^{-3}]$. Substituting the given values, $\mathbf{F}_3 = (8\hat{\imath} + 15\hat{\jmath})$ N, $\mathbf{r}_1 = 2\hat{\jmath}$ m, $\mathbf{r}_2 = 3\hat{\imath}$ m, and $\mathbf{r}_3 = (4\hat{\imath} + 3\hat{\jmath})$ m, we find $(8\hat{\imath} + 15\hat{\jmath})$ N·m$^2 = kq_3\times [q_1(4\hat{\imath}+\hat{\jmath})(4^2 + 1^2)^{-3/2} + q_2(\hat{\imath} + 3\hat{\jmath})(1^2 + 3^2)^{-3/2}]$. Equating x and y components, we get $8 \text{ N·m}^2/kq_3 = 4q_1 17^{-3/2} + q_2 10^{-3/2}$ and $15 \text{ N·m}^2/kq_3 = q_1 17^{-3/2} + 3q_2 10^{-3/2}$. Dividing these equations and solving for q_2 in terms of $q_1 = 55$ μC, we find $q_2 = (10/17)^{3/2}\times (52/9)q_1 = 143$ μC. Substituting this into either component equation, we get $q_3 = (8 \text{ N·m}^2/k)\times [4q_1 17^{-3/2} + q_2 10^{-3/2}]^{-1} = (15 \text{ N·m}^2/k)\,[q_1 17^{-3/2} + 3q_2 10^{-3/2}]^{-1} = 116$ μC.

Section 23-4: The Electric Field

Problem

25. An electron placed in an electric field experiences a 6.1×10^{-10} N electric force. What is the field strength?

Solution

From Equation 23-3a, $E = F/e = 6.1\times10^{-10}$ N÷ 1.6×10^{-19} C $= 3.81\times10^9$ N/C. (The field strength is the magnitude of the field.)

Problem

27. A 68-nC charge experiences a 150-mN force in a certain electric field. Find (a) the field strength and (b) the force that a 35-μC charge would experience in the same field.

Solution

Equations 23-3a and b give (a) $E = 150$ mN/68 nC $=$ 2.21 MN/C, and (b) $F = (35$ μC$)(2.21$ MN/C$) =$ 77.2 N.

Problem

29. The electron in a hydrogen atom is 0.0529 nm from the proton. What is the proton's electric field strength at this distance?

Solution

The proton in a hydrogen atom behaves like a point charge, for an electron one Bohr radius away (see solution to Problem 7), so Equation 23-4 gives $E = ke/a_0^2 = (9\times10^9 \text{ N·m}^2/\text{C}^2)(1.6\times10^{-19} \text{ C})\div (5.29\times10^{-11} \text{ m})^2 = 5.15\times10^{11}$ N/C.

Section 23-5: Electric Fields of Charge Distributions

Problem

31. In Fig. 23-40, point P is midway between the two charges. Find the electric field in the plane of the page (a) 5.0 cm directly above P, (b) 5.0 cm directly to the right of P, and (c) at P.

Solution

Take the origin of x-y coordinates at the midpoint, as indicated, and use Equation 23-5. Let $\mathbf{r}_\pm = \pm(2.5\text{ cm})\hat{\jmath}$ denote the positions of the charges, and $\mathbf{r}$ that of the field point. A unit vector from one charge to the field point is $(\mathbf{r}-\mathbf{r}_\pm)/|\mathbf{r}-\mathbf{r}_\pm|$, so the spacial factors in Coulomb's law are $\hat{\mathbf{r}}_i/r_i^2 = \mathbf{r}_i/r_i^3 = (\mathbf{r}-\mathbf{r}_\pm)/|\mathbf{r}-\mathbf{r}_\pm|^3$. (a) For $\mathbf{r} = (5.0\text{ cm})\hat{\jmath}, \mathbf{r}_1 = \mathbf{r}-\mathbf{r}_+ = (5.0\text{ cm})\hat{\jmath} - (2.5\text{ cm})\hat{\jmath} = (2.5\text{ cm})\hat{\jmath}$, and $\mathbf{r}_2 = \mathbf{r}_2 = \mathbf{r}-\mathbf{r}_- = (7.5\text{ cm})\hat{\jmath}$. Then

$$\mathbf{E} = k\left(\frac{q_1\mathbf{r}_1}{r_1^3} + \frac{q_2\mathbf{r}_2}{r_2^3}\right)$$
$$= \left(9\times10^9\frac{\text{N}\cdot\text{m}^2}{\text{C}^2}\right)(2\ \mu\text{C})\left[\frac{\hat{\jmath}}{(2.5\text{ cm})^2} - \frac{\hat{\jmath}}{(7.5\text{ cm})^2}\right]$$
$$= (25.6\text{ MN/C})\hat{\jmath}.$$

(b) For $\mathbf{r} = (5.0\text{ cm})\hat{\imath}$,

$$\mathbf{E} = \left(9\times10^9\frac{\text{N}\cdot\text{m}^2}{\text{C}^2}\right)\left(\frac{2\ \mu\text{C}}{\text{cm}^2}\right)\left[\frac{(5.0\hat{\imath}-2.5\hat{\jmath})}{(5.0^2+(-2.5)^2)^{3/2}} - \frac{(5.0\hat{\imath}+2.5\hat{\jmath})}{(5.0^2+2.5^2)^{3/2}}\right] = -(5.15\text{ MN/C})\hat{\jmath}.$$

(c) For $\mathbf{r} = 0$,

$$\mathbf{E} = \left(9\times10^9\frac{\text{N}\cdot\text{m}^2}{\text{C}^2}\right)\left(\frac{2\ \mu\text{C}}{\text{cm}^2}\right)\left[\frac{-\hat{\jmath}}{(2.5)^2} - \frac{\hat{\jmath}}{(2.5)^2}\right]$$
$$= -(57.6\text{ MN/C})\hat{\jmath}.$$

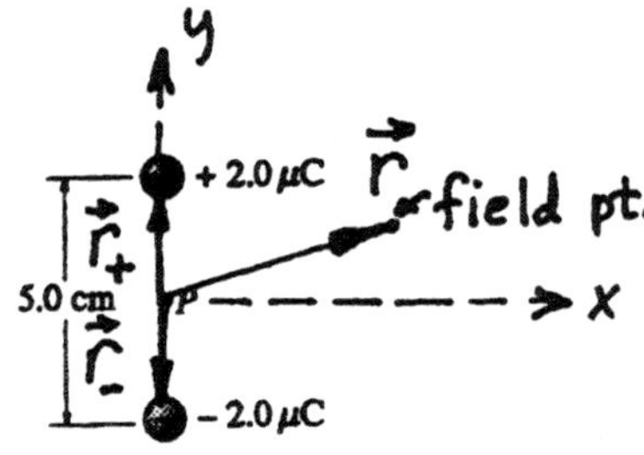

FIGURE 23-40 Problem 31 Solution.

Problem

33. A proton is at the origin and an ion is at $x = 5.0$ nm. If the electric field is zero at $x = -5$ nm, what is the charge on the ion?

Solution

The proton, charge e, is at $\mathbf{r}_p = 0$, and the ion, charge q, is at $\mathbf{r}_I = 5\hat{\imath}$ nm. The field at point $\mathbf{r} = -5\hat{\imath}$ nm is given by Equation 23-5, with spacial factors written as in the solutions to Problems 15 or 31:

$$\mathbf{E}(\mathbf{r}) = \sum_i kq_i\frac{(\mathbf{r}-\mathbf{r}_i)}{|\mathbf{r}-\mathbf{r}_i|^3}$$
$$= ke\frac{(-5\hat{\imath}\text{ nm})}{(5\text{ nm})^3} + kq\frac{(-5\hat{\imath}\text{ nm} - 5\hat{\imath}\text{ nm})}{(10\text{ nm})^3}.$$

Therefore, $\mathbf{E} = 0$ implies $2q/(10)^3 = -e/(5)^3$, or $q = -4e$. (Note how we used the general expression for the electric field, at position $\mathbf{r}$, due to a distribution of static point charges at positions $\mathbf{r}_i$.)

Problem

35. (a) Find an expression for the electric field on the y-axis due to the two charges q in Fig. 23-11. (b) At what point is the field on the y-axis a maximum?

Solution

(a) The electric field is the force per unit charge, so Example 23-2 shows that $\mathbf{E}(y) = 2kqy\hat{\jmath}(a^2+y^2)^{-3/2}$. (b) The magnitude of the field, a positive function, is zero for $y = 0$ and $y = \infty$, hence it has a maximum in between. Setting the derivative equal to zero, we find $0 = (a^2+y^2)^{-3/2} - \frac{3}{2}y(a^2+y^2)^{-5/2}(2y)$, or $a^2+y^2-3y^2 = 0$. Thus, the field strength maxima are at $y = \pm a/\sqrt{2}$ (the directions at these points, of course, are opposite, by symmetry).

Problem

37. A dipole lies on the y-axis, and consists of an electron at $y = 0.60$ nm and a proton at $y = -0.60$ nm. Find the electric field (a) midway between the two charges, (b) at the point $x = 2.0$ nm, $y = 0$, and (c) at the point $x = -20$ nm, $y = 0$.

Solution

We can use the result of Example 23-6, with y replaced by x, and x by $-y$ (or equivalently, $\hat{\jmath}$ by $\hat{\imath}$, and $\hat{\imath}$ by $-\hat{\jmath}$). Then $\mathbf{E}(x) = 2kqa\,\hat{\jmath}(a^2+x^2)^{-3/2}$, where $q = e = 1.6\times10^{-19}$ C and $a = 0.6$ nm. (Look at Fig. 23-18 rotated 90° CW.) The constant $2kq = 2(9\times10^9\text{ N}\cdot\text{m}^2/\text{C}^2)(1.6\times10^{-19}\text{ C}) = (2.88\text{ GN/C})\times(\text{nm})^2$. (a) At $x = 0$, $\mathbf{E}(0) = 2kq\hat{\jmath}/a^2 = (2.88\text{ GN/C})\hat{\jmath}/(0.6)^2 = (8.00\text{ GN/C})\hat{\jmath}$. (b) For $x = 2$ nm, $\mathbf{E} = (2.88\text{ GN/C})\hat{\jmath}(0.6)(0.6^2+2^2)^{-3/2} = (190\text{ MN/C})\hat{\jmath}$. (c) At $x = 20$ nm, $\mathbf{E} = (2.88\text{ GN/C})\hat{\jmath}\times(0.6)(0.6^2+20^2)^{-3/2} = (216\text{ kN/C})\hat{\jmath}$.

Problem

39. The dipole moment of the water molecule is 6.2×10^{-30} C·m. What would be the separation distance if the molecule consisted of charges $\pm e$? (The effective charge is actually less because electrons are shared by the oxygen and hydrogen atoms.)

Solution

The distance separating the charges of a dipole is $d = p/q = 6.2\times10^{-30}\text{ C·m}/1.6\times10^{-19}\text{ C} = 38.8$ pm.

Problem

41. Three charges form an equilateral triangle of side a. At one vertex is a charge $+2q$; at the other two vertices are charges $-q$. The triangle is oriented with the charge $2q$ on the positive x-axis and both charges $-q$ on the y-axis. (a) Find an expression for the electric field on the x-axis, in the approximation $x \gg a$. (b) Compare with Equation 23-7b to show that your result in (a) is a dipole field, and give an expression for the magnitude of the triangle's dipole moment.

Solution

(a) With the charges positioned as shown, the electric field on the positive x-axis, due to the two negative charges at $(0, \pm a/2)$, matches the field found in Example 23-2 (replace q with $-q$, a with $a/2$, y with x, and $\mathbf{\hat{j}}$ with $\mathbf{\hat{i}}$): $\mathbf{F}/Q = 2k(-q) \times (x^2 + a^2/4)^{-3/2}\mathbf{\hat{i}}$. For the positive charge at $(\sqrt{3}a/2, 0)$, the electric field on the x-axis to the right $(x > \sqrt{3}a/2)$ is just $k(2q)\times(x - \sqrt{3}a/2)^{-2}\mathbf{\hat{i}}$ (the unit vector in Equation 23-5 is $\mathbf{\hat{i}}$ and the distance is $x - \sqrt{3}a/2$). The total field is the sum of these, $\mathbf{E}(x) = 2kq\mathbf{\hat{i}}[(x - \sqrt{3}a/2)^{-2} - x(x^2 + a^2/4)^{-3/2}]$. For $x \gg a$, one can use the binomial approximation (see Appendix A): $(x - \sqrt{3}a/2)^{-2} = x^{-2}(1 + \sqrt{3}a/x + \cdots)$ and $x(x^2 + a^2/4)^{-3/2} = x^{-2}(1 + \cdots)$, where $\cdots$ indicates terms of order a^2/x^2 or higher. Therefore, $\mathbf{E}(x) = 2kq\mathbf{\hat{i}}x^{-2}[1 + \sqrt{3}a/x + \cdots - (1 + \cdots)] \simeq 2\sqrt{3}kqax^{-3}\mathbf{\hat{i}}$. (b) This field is the same as Equation 23-7b, with $p = \sqrt{3}qa$. (In general, the dipole moment for a distribution of point charges is $|\sum \mathbf{r}_i q_i|$. Also note that the field in part (a) can also be found from Equation 23-5, with $\mathbf{\hat{r}}_i/r_i^2 = (\mathbf{r} - \mathbf{r}_i)\div|\mathbf{r} - \mathbf{r}_i|^3$ as in the solution to Problem 15.)

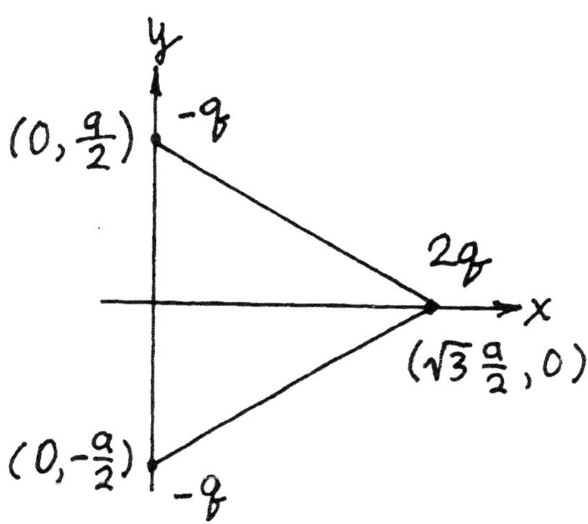

Problem 41 Solution.

Problem

43. A 30-cm-long rod carries a charge of 80 μC spread uniformly over its length. Find the electric field strength on the rod axis, 45 cm from the end of the rod.

Solution

Applying the result of Example 23-7, at a distance $a = 0.45$ m from the near end of the rod, we get $E = kQ/a(a + \ell) = (9\times10^9\text{ N·m}^2/\text{C}^2)(80\ \mu\text{C})\div(0.45\text{ m})(0.45\text{ m} + 0.30\text{ m}) = 2.13$ MN/C.

Problem

45. A thin rod of length ℓ has its left end at the origin and its right end at the $x = \ell$. It carries a line charge density given by $\lambda = \lambda_0(x^2/\ell^2)\sin(\pi x/\ell)$, where λ_0 is a constant. Find the electric field strength at the origin.

Solution

The electric field at the origin, due to a small element of charge, $dq = \lambda\ dx$, located at position x, is $d\mathbf{E} =$

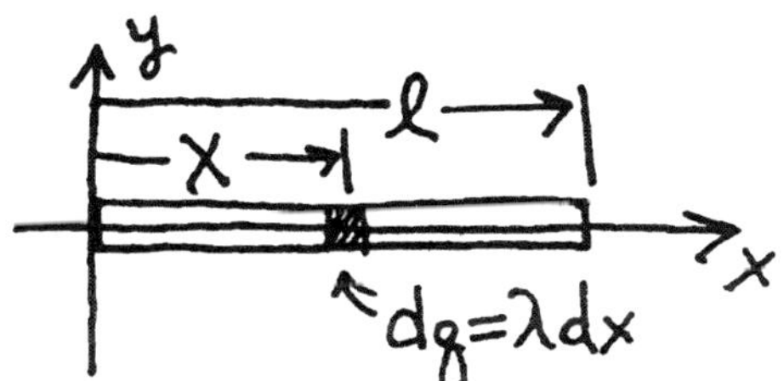

Problem 45 Solution.

$-\hat{\imath}k\lambda\, dx/x^2$. Using $\lambda/x^2 = (\lambda_0/\ell^2)\sin(\pi x/\ell)$ and integrating from $x = 0$ to $x = \ell$, we find

$$\begin{aligned}\mathbf{E} &= \int_0^\ell -\hat{\imath}\left(\frac{k\lambda_0}{\ell^2}\right)\sin\left(\frac{\pi x}{\ell}\right)\,dx\\ &= \hat{\imath}\frac{k\lambda_0}{\ell^2}\left|\frac{\ell}{\pi}\cos\left(\frac{\pi x}{\ell}\right)\right|_0^\ell\\ &= \hat{\imath}(k\lambda_0/\ell\pi)(-1-1) = -2\hat{\imath}k\lambda_0/\pi\ell.\end{aligned}$$

Problem

47. A uniformly charged ring is 1.0 cm in radius. The electric field on the axis 2.0 cm from the center of the ring has magnitude 2.2 MN/C and points toward the ring center. Find the charge on the ring.

Solution

From Example 23-8, the electric field on the axis of a uniformly charged ring is $kQx(x^2 + a^2)^{-3/2}$, where x is positive away from the center of the ring. For the given ring in this problem, $-2.2\text{ MN/C} = kQ(2\text{ cm})\times(4\text{ cm}^2 + 1\text{ cm}^2)^{-3/2}$, or $Q = (-2.2\text{ MN/C})\times(5.59\text{ cm}^2)/(9\times10^9\text{ N}\cdot\text{m}^2/\text{C}) = -0.137\ \mu\text{C}$.

Problem

49. Use the result of the preceding problem to show that the field of an *infinite*, uniformly charged flat sheet is $2\pi k\sigma$, where σ is the surface charge density. Note that this result is independent of distance from the sheet.

Solution

An infinite flat sheet is the same as an infinite flat disk (as long as the dimensions are infinite in all directions, the shape is irrelevant). Thus, we can find the magnitude of the electric field from a uniformly changed infinite flat sheet by letting $R \to \infty$ in the result of the previous problem. Then, the limit of the second term is zero, and the magnitude is constant, $E = 2\pi k\sigma$. (The direction is perpendicularly away from (towards) the sheet for positive (negative) σ.)

Problem

51. The electric field 22 cm from a long wire carrying a uniform line charge density is 1.9 kN/C. What will be the field strength 38 cm from the wire?

Solution

For a very long wire ($\ell \gg 38$ cm), Example 23-9 shows that the magnitude of the radial electric field falls off like $1/r$. Therefore, $E(38\text{ cm})/E(22\text{ cm}) = 22\text{ cm}\div 38\text{ cm}$; or $E(38\text{ cm}) = (22/38)1.9\text{ kN/C} = 1.10\text{ kN/C}$.

Problem

53. A straight wire 10 m long carries 25 μC distributed uniformly over its length. (a) What is the line charge density on the wire? Find the electric field strength (b) 15 cm from the wire axis, not near either end and (c) 350 m from the wire. Make suitable approximations in both cases.

Solution

(a) For a uniformly charged wire, $\lambda = Q/\ell = 2.5\ \mu\text{C/m}$. (b) Since $r = 15\text{ cm} \ll 10\text{ m} = \ell$ and the field point is far from either end, we may regard the wire as approximately infinite. Then Example 23-9 gives $E_r = 2k\lambda/r = (2\times 9\times10^9\text{ N}\cdot\text{m}^2/\text{C}^2)(2.5\ \mu\text{C/m})\div(0.15\text{ m}) = 300\text{ kN/C}$. (c) At $r = 350\text{ m} \gg 10\text{ m} = \ell$, the wire behaves approximately like a point charge, so the field strength is $kQ/r^2 = (9\times10^9\times 25\times10^{-6}\text{ N}\cdot\text{m}^2/\text{C})/(350\text{ m})^2 = 1.84\text{ N/C}$.

Section 23-6: Matter in Electric Fields

Problem

55. In this famous 1909 experiment that demonstrated quantization of electric charge, R. A. Millikan suspended small oil drops in an electric field. With a field strength of 20 MN/C, what mass drop can be suspended when the drop carries a net charge of 10 elementary charges?

Solution

In equilibrium under the gravitational and electrostatic forces, $mg = qE$, or $m = (10\times 1.6\times10^{-19}\text{ C})\times(2\times10^7\text{ N/C})/(9.8\text{ m/s}^2) = 3.27\times10^{-12}$ kg. (Because this is so small, the size of such a drop may be better appreciated in terms of its radius, $R = (3m/4\pi\rho_{\text{oil}})^{1/3}$. Millikan used oil of density 0.9199 g/cm^3, so $R = 9.46\ \mu$m for this drop.)

Problem

57. A proton moving to the right at 3.8×10^5 m/s enters a region where a 56 kN/C electric field points to the left. (a) How far will the proton get before its speed reaches zero? (b) Describe its subsequent motion.

Solution

(a) Choose the x-axis to the right, in the direction of the proton, so that the electric field is negative to the left. If the Coulomb force on the proton is the only important one, the acceleration is $a_x = e(-E)/m$. Equation 2-11, with $v_{0x} = 3.8\times10^5$ m/s and $v_x = 0$, gives a maximum penetration into the field region of

$x - x_0 = -v_{ox}^2/2a_x = mv_{ox}^2/2eE =$

$$\frac{(1.67\times10^{-27}\text{ kg})(3.8\times10^5\text{ m/s})^2}{2(1.6\times10^{-19}\text{ C})(56\times10^3\text{ N/C})} = 1.35\text{ cm}.$$

(b) The proton then moves to the left, with the same constant acceleration in the field region, until it exits with the initial velocity reversed.

Problem

59. An ink-jet printer works by "steering" charged ink drops to the right place on the page by passing moving drops through a uniform electric field that deflects them by the appropriate amount. Figure 23-47 shows an ink drop approaching the field region, which has length ℓ and width d between the charged plates that establish the field. Find an expression for the minimum speed a drop with mass m and charge q must have if it is to get through the region without hitting either plate.

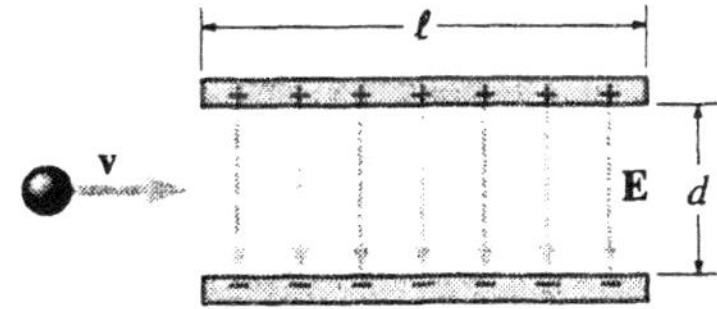

FIGURE 23-47 Problem 59.

Solution

If they enter the field region midway, moving horizontally, the maximum vertical deflection, during the transit time $t = \ell/v$, can be $d/2$, for the ink drops to pass through. Thus, $y = \frac{1}{2}at^2 = \frac{1}{2}(qE/m)(\ell/v)^2 < \frac{1}{2}d$, or $v > \ell\sqrt{qE/md}$.

Problem

61. An electron is moving in a circular path around a long, uniformly charged wire carrying 2.5 nC/m. What is the electron's speed?

Solution

The electric field of the wire is radial and falls off like $1/r$ (see Example 23-9). For an attractive force (negative electron encircling a positively charged wire), this is the same dependance as the centripetal acceleration. For circular motion around the wire, the Coulomb force provides the electron's centripetal acceleration, or $-eE/m = -2ke\lambda/mr = -v^2/r$. Thus, $v = \sqrt{2ke\lambda/m} = [2(9\times10^9\text{ N}\cdot\text{m}^2/\text{C}^2)(1.6\times10^{-19}\text{ C})\times(2.5\times10^{-9}\text{ C/m})/(9.11\times10^{-31}\text{ kg})]^{1/2} = 2.81\text{ Mm/s}$.

Problem

63. What is the line charge density on a long wire if a 6.8-μg particle carrying 2.1 nC describes a circular orbit about the wire with speed 280 m/s?

Solution

The solution to Problem 61 reveals that $\lambda = -mv^2 \div 2kq = -(6.8\times10^{-9}\text{ kg})(280\text{ m/s})^2/2(9\times10^9\text{ N}\cdot\text{m}^2/\text{C}^2)\times(2.1\times10^{-9}\text{ C}) = -14.1\ \mu\text{C/m}$. (In this case, the force on a positively charged orbiting particle is attractive for a wire with negative linear charge density.)

Problem

65. A dipole with dipole moment 1.5 nC·m is oriented at 30° to a 4.0-MN/C electric field. (a) What is the magnitude of the torque on the dipole? (b) How much work is required to rotate the dipole until it's antiparallel to the field?

Solution

(a) The torque on an electric dipole in an external electric field is given by Equation 23-11; $\tau = |\mathbf{p}\times\mathbf{E}| = pE\sin\theta = (1.5\text{ nC}\cdot\text{m})(4.0\text{ MN/C})\sin 30° = 3.0\text{ mN}\cdot\text{m}$. (b) The work done against just the electric force is equal to the change in the dipole's potential energy (Equation 23-12); $W = \Delta U = (-\mathbf{p}\cdot\mathbf{E})_f - (-\mathbf{p}\cdot\mathbf{E})_i = pE(\cos 30° - \cos 180°) = (1.5\text{ nC}\cdot\text{m})(4.0\text{ MN/C})\times(1.866) = 11.2\text{ mJ}$.

Problem

67. Two identical dipoles, each of charge q and separation a, are a distance x apart as shown in Fig.23-49. By considering forces between pairs of charges in the different dipoles, calculate the net force between the dipoles. (a) Show that, in the limit $a \ll x$, the force has magnitude $6kp^2/x^4$, where $p = qa$ is the dipole moment. (b) Is the force attractive or repulsive?

Solution

All the forces are along the same line, so take the origin at the center of the left-hand dipole and the positive x-axis in the direction of the right-hand dipole in Fig. 23-49. The right-hand dipole has charges $+q$ at $x + a/2$, $-q$ at $x - a/2$, each of which experiences a force from both charges of the left-hand dipole, which are $+q$ at $a/2$ and $-q$ at $-a/2$. (There are forces between four pairs of changes.) The Coulomb force on a charge in the right-hand dipole, due to one in the left-hand one, is $kq_rq_\ell(x_r - x_\ell)\hat{\imath}/|x_r - x_\ell|^3$ (see solution to Problem 15), so the total force on the right-hand dipole is

$$F_x = kq^2\hat{\imath}\left[\frac{1}{x^2} - \frac{1}{(x+a)^2} - \frac{1}{(x-a)^2} + \frac{1}{x^2}\right] = -\frac{2kq^2a^2(3x^2 - a^2)}{x^2(x^2 - a^2)^2}\hat{\imath}.$$

(a) In the limit $a \ll x$, $F_x \to -2kq^2a^2(3x^2)\hat{\imath}/x^6 = -6kq^2a^2\hat{\imath}/x^4 = -6kp^2\hat{\imath}/x^4$, where $p = qa$ is the dipole moment of both dipoles. (b) The force on the right-hand dipole is in the negative x direction, indicating an attractive force.

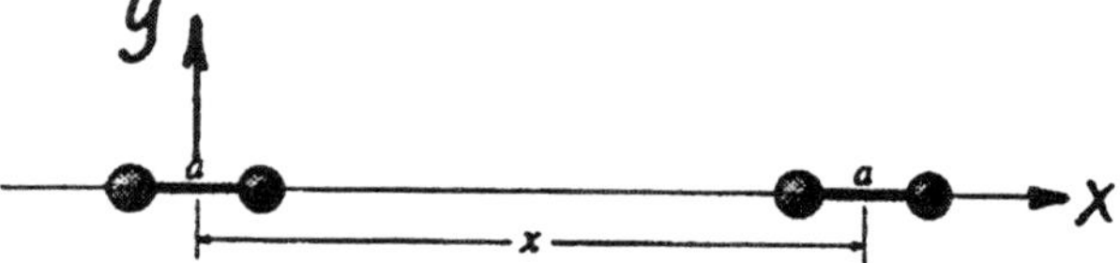

FIGURE 23-49 Problem 67 Solution.

Paired Problems

Problem

69. An electron is at the origin and an ion with charge $+5e$ is at $x = 10$ nm. Find a point where the electric field is zero.

Solution

The electron's field is directed toward the electron (a negative charge) and the ion's field is directed away from the ion (a positive charge). Therefore, the fields can cancel only at points on the negative x -axis ($x < 0$), since the directions are opposite there and the smaller charge is closer. The field from one point charge is $\mathbf{E}_q(x) = kq\hat{\imath}(x - x_q)/|x - x_q|^3$, where $q = -e$, $x_q = 0$ for the electron, and $q = 5e$, $x_q = 10$ nm for the ion. The total field is zero when $0 = k[(-e)x\times |x|^{-3} + 5e(x - 10 \text{ nm})\,|x - 10 \text{ nm}|^{-3}]$. (See note to solution of Problem 33.) Since $x < 0$, $|x| = -x$ and $|x - 10 \text{ nm}| = 10 \text{ nm} - x$, so this implies $x^{-2} - 5\times (10 \text{ nm} - x)^{-2} = 0$, or $4x^2 + 2(10 \text{ nm})x - (10 \text{ nm})^2 = 0$. The negative solution to this quadratic is

$$x = \frac{[-10 \text{ nm} - \sqrt{(10 \text{ nm})^2 + 4(10 \text{ nm})^2}]}{4} = -2.5 \text{ nm}(1 + \sqrt{5}) = -8.09 \text{ nm}.$$

Problem

71. A thin rod of length ℓ has its left end at $x = -\ell$ and its right end at the origin. It carries a line charge density given by

$$\lambda = \lambda_0 \frac{x^2}{\ell^2},$$

where λ_0 is a constant. Find the electric field at the origin.

Solution

The electric field at the origin, due to an element of charge $dq = \lambda\, dx$, located at x, where $-\ell \le x \le 0$, is $d\mathbf{E} = k\lambda\hat{\imath}\, dx/x^2 = k(\lambda_0/\ell^2)\hat{\imath}\, dx$. The total field is the integral of this over the rod,

$$\mathbf{E}(0) = \int_{-\ell}^{0} (k\lambda_0/\ell^2)\hat{\imath}\, dx = (k\lambda_0/\ell^2)\hat{\imath}[0 - (-\ell)] = (k\lambda_0/\ell)\hat{\imath}.$$

Problem

73. A thin, flexible rod carrying charge Q spread uniformly over its length is bent into a quarter circle of radius a, as shown in Fig. 23-51a. Find the electric field strength at the point P, which is the center of the circle. *Hint:* Consult Problem 50.

Solution

It should be clear from the symmetry that the electric field is along the radius bisecting the arc, so take this as the x-axis, with P at the origin, and θ as defined in Fig. 23-45. The electric field at P, from each charge element, $dq = \lambda\, d\ell = \lambda a\, d\theta$, at θ, has the same magnitude, $dE = k\, dq/a^2 = k\lambda\, d\theta/a$, but direction $\hat{\mathbf{r}} = \hat{\imath}\sin\theta - \hat{\jmath}\cos\theta$, as sketched. $\lambda = Q/\ell$ is constant, and the arc extends from $\theta_0 = 45°$ to $\pi - \theta_0 = 135°$, so the total field at P is the integral of $dE\hat{\mathbf{r}}$ from θ_0 to $\pi - \theta_0$:

$$\begin{aligned}\mathbf{E}(0) &= (k\lambda/a)\int_{\theta_0}^{\pi-\theta_0} (\hat{\imath} \sin\theta - \hat{\jmath} \cos\theta)\, d\theta \\ &= (k\lambda/a)|-\hat{\imath} \cos\theta - \hat{\jmath} \sin\theta|_{\theta_0}^{\pi-\theta_0} \\ &= (k\lambda/a)2\hat{\imath} \cos\theta_0 = \sqrt{2}\, k\lambda\, \hat{\imath}/a.\end{aligned}$$

Here, we used $\sin\theta_0 = \sin(\pi - \theta_0)$, $\cos\theta_0 = -\cos(\pi - \theta_0)$, and $\theta_0 = 45°$. In terms of the total charge, $\lambda = Q/\ell = Q/(\frac{1}{2}\pi a)$, so $\mathbf{E}(0) = 2\sqrt{2}\, kQ\hat{\imath} \div \pi a^2$. [Note: in general, $\ell = (\pi - 2\theta_0)a$.]

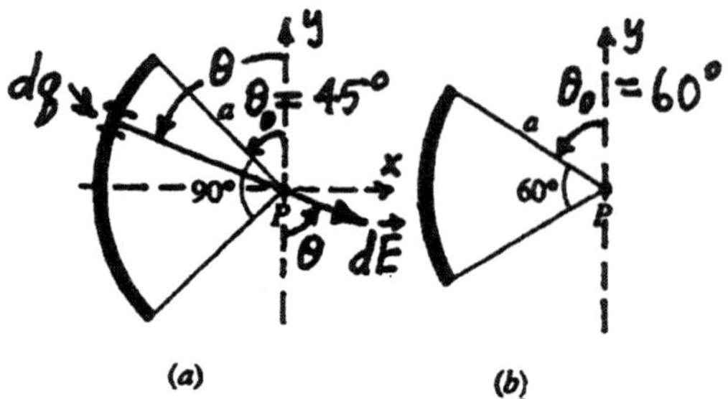

FIGURE 23-51 Problems 73 and 74 Solution.

Problem

75. Ink-jet printers work by deflecting moving ink droplets with an electric field so they hit the right place on the paper. Droplets in a particular printer have mass 1.1×10^{-10} kg, charge 2.1 pC, speed

12 m/s, and pass through a uniform 97-kN/C electric field in order to be deflected through a 10° angle. What is the length of the field region?

Solution

Suppose the ink droplets enter the field region perpendicular to the field, as in the geometry of Example 23-10. Then the analysis of that example shows that $v_y/v_x = \tan\theta = qE_y\ \Delta x/mv_x^2$, so $\Delta x = mv_x^2\tan\theta/qE_y = (0.11\ \mu\text{g})(12\ \text{m/s})^2\tan 10°\div (2.1\ \text{pC})(97\ \text{kN/C}) = 1.37\ \text{cm}$.

Supplementary Problems

Problem

77. A spring of spring constant 100 N/m is stretched 10 cm beyond its 90-cm equilibrium length. If you want to keep it stretched by attaching equal electric charges to the opposite ends, what magnitude of charge should you use?

Solution

The repulsive force between like charges, $kq^2/r^2\times$ ($r = 90\ \text{cm} + 10\ \text{cm} = 1\ \text{m}$), must balance the spring force, k_sx ($x = 10$ cm is the stretch and k_s is the spring constant). Thus,

$$\begin{aligned} q &= \pm\sqrt{r^2k_sx/k} \\ &= \pm[(100\ \text{N/m})(1\ \text{m})^2(0.1\ \text{m})/(9\times10^9\ \text{N}\cdot\text{m}^2/\text{C}^2)]^{1/2} \\ &= \pm 33.3\ \mu\text{C}. \end{aligned}$$

Problem

79. A charge $-q$ and a charge $\frac{4}{9}q$ are located a distance a apart, as shown in Fig. 23-53. Where would you place a third charge so that all three are in static equilibrium? What should be the sign and magnitude of the third charge?

Solution

Because of the vector nature of the forces, the third charge, Q, must be placed along the line joining the other two, as in Example 23-3. Q cannot go to the left of $-q$, since the magnitude of the force on it due to $-q$ would always be greater than that due to $\frac{4}{9}q$. It cannot go between $-q$ and $\frac{4}{9}q$, since the forces on it would always be in the same direction. Thus, Q must go to the right of $\frac{4}{9}q$, as shown (x-axis to the right with origin at $\frac{4}{9}q$). The net force on each charge must be zero, so, for Q: $0 = kQ\left(\frac{4}{9}q\right)/x^2 + kQ(-q)/(x+a)^2$, or $\frac{4}{9}(x+a)^2 = x^2$. For $\frac{4}{9}q$: $0 = k\left(\frac{4}{9}q\right)(-q)/a^2 - k\times \left(\frac{4}{9}q\right)Q/x^2$, or $-qx^2 = Qa^2$. (The equation for the force on the third charge follows from the equations for the other two plus Newton's third law.) The solution of these two equations (for $x > 0$) is $x = 2a$ and $Q = -4q$. (The equilibrium is unstable. A slight displacement of the positive charge to the right, for example, would cause it to be attracted more strongly to the right.)

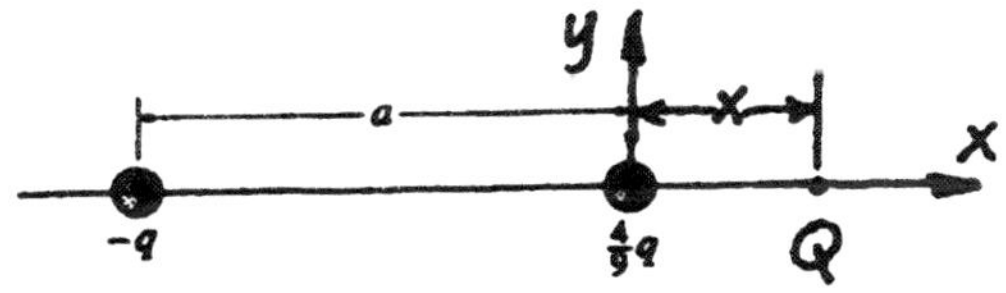

FIGURE 23-53 Problem 79 Solution.

Problem

81. A 3.8-g particle with a 4.0-μC charge experiences a downward force of 0.24 N in a uniform electric field. Find the electric field, assuming that the gravitational force is *not* negligible.

Solution

If gravity and Coulomb forces both act, then $\mathbf{F}_{\text{net}} = -mg\hat{\mathbf{j}} + q\mathbf{E} = -(0.24\ \text{N})\hat{\mathbf{j}}$, where $\hat{\mathbf{j}}$ is upward. Thus, $\mathbf{E} = (mg - 0.24\ \text{N})\hat{\mathbf{j}}/q = [(3.8\times10^{-3}\ \text{kg})(9.8\ \text{m/s}^2) - 0.24\ \text{N}]\hat{\mathbf{j}}/(4.0\ \mu\text{C}) = -50.7\hat{\mathbf{j}}\ \text{kN/C}$.

Problem

83. The electric field on the axis of a uniformly charged ring has magnitude 380 kN/C at a point 5.0 cm from the ring center. The magnitude 15 cm from the center is 160 kN/C; in both cases the field points away from the ring. Find the radius and charge of the ring.

Solution

The electric field on the axis of a uniformly charged ring is calculated in Example 23-8, so the data given in the question imply $380\ \text{kN/C} = kQ(5\ \text{cm})\times [(5\ \text{cm})^2 + a^2]^{-3/2}$, and $160\ \text{kN/C} = kQ(15\ \text{cm})\times [(15\ \text{cm})^2 + a^2]^{-3/2}$. Dividing these two equations and taking the $\frac{2}{3}$ root we get

$$\left(\frac{380\times15}{160\times5}\right)^{2/3} = 3.70 = \frac{(15\ \text{cm})^2 + a^2}{(5\ \text{cm})^2 + a^2},$$

which when solved for the radius, gives

$$a = \sqrt{[(15\ \text{cm})^2 - (3.70)(5\ \text{cm})^2]/2.70} = 7.00\ \text{cm}.$$

Substituting for a in either of the field equations allows us to find

$$Q = \frac{(380\ \text{kN/C})[(5\ \text{cm})^2 + (7\ \text{cm})^2]^{3/2}}{(9\times10^9\ \text{N}\cdot\text{m}^2/\text{C}^3)(5\ \text{cm})} = 538\ \text{nC}.$$

Problem

85. A molecule with dipole moment p is located a distance r from a proton, oriented with its dipole moment vector $\mathbf{p}$ as shown in Fig. 23-54. (a) Use Equation 23-7b to find the force the molecule exerts on the proton. (b) Now find the net force on the molecule in the proton's nonuniform electric field by considering that the molecule consists of two opposite charges $\pm q$, separated by a distance d such that $qd = p$. Take the limit as d becomes very small compared with r, and show that the resulting force has the same magnitude as that of part (a), as required by Newton's third law.

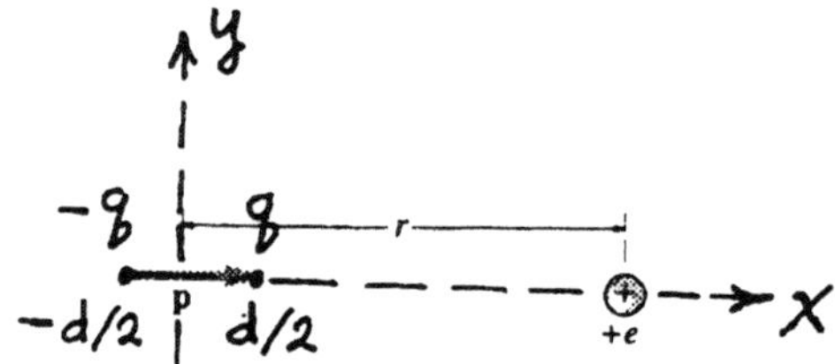

FIGURE 23-54 Problem 85.

Solution

Take the origin at the molecule, with the x-axis parallel to its dipole moment, as shown on Fig. 23-54. (a) Equation 23-7b gives the dipole's electric field at the position of the proton, so the force on the latter is $\mathbf{F}_p = e\mathbf{E}_{\text{dip}} = e(2kpr^{-3}\hat{\mathbf{i}}) = 2kepr^{-3}\hat{\mathbf{i}}$. (b) The electric field of the proton, for points on the x-axis to its left ($x < r$) is $\mathbf{E}_p(x) = -ke(r - x)^{-2}\hat{\mathbf{i}}$ (see Equation 23-4). If the molecular dipole is considered to consist of charges $\pm q$ located at $x = \pm d/2$, where $p = qd$, then the force on it is $\mathbf{F}_{\text{dip}} = q\mathbf{E}_p(d/2) - q\mathbf{E}_p(-d/2) = -keq\,\hat{\mathbf{i}}[(r - d/2)^{-2} - (r + d/2)^{-2}] = -2kepr \times (r^2 - d^2/4)^{-2}\hat{\mathbf{i}}$. In the limit $d \ll r$, this becomes $\mathbf{F}_{\text{dip}} = -2kepr^{-3}\hat{\mathbf{i}} = -\mathbf{F}_p$, in agreement with the result of part (a) and Newton's third law.

Problem

87. Derive Equation 23-9 in Example 23-9 by making θ the integration variable, then evaluating the resulting integral.

Solution

In Fig. 23-23, $r = (x^2 + y^2)^{1/2} = y/\sin\theta$, and $\cot\theta = x/y$. Differentiating the latter with respect to x, we get $-\csc^2\theta\, d\theta = dx/y$, which relates dx to $d\theta$. (Extend the results in Appendix B, and recall that cosecant = sin e^{-1}.) The limits $x = -\infty$ to ∞ correspond to $\theta = \pi$ to 0 (since when x is negative so is $\cot\theta$), thus the integral in Example 23-9 leading to Equation 23-9 becomes

$$\begin{aligned} E &= k\lambda y \int_{-\infty}^{\infty} \frac{dx}{(x^2 + y^2)^{3/2}} \\ &= k\lambda y \int_{\pi}^{0} \frac{-y\csc^2\theta\, d\theta}{(y/\sin\theta)^3} = \frac{k\lambda}{y}\int_0^{\pi} \sin\theta\, d\theta \\ &= (k\lambda/y)[-\cos\pi + \cos 0] = 2k\lambda/y \end{aligned}$$

as before.

CHAPTER 24 GAUSS'S LAW

ActivPhysics can help with these problems: Activities 11.4–11.6

Section 24-1: Electric Field Lines

Problem

1. What is the net charge shown in Fig. 24-39? The magnitude of the middle charge is 3 μC.

Solution

The number of lines of force emanating from (or terminating on) the positive (or negative) charges is the same (14 in Fig. 24-39), so the middle charge is $-3\ \mu$C and the outer ones are $+3\ \mu$C. The net charge shown is therefore $3 + 3 - 3 = 3\ \mu$C. This is reflected by the fact that 14 lines emerge from the boundary of the figure.

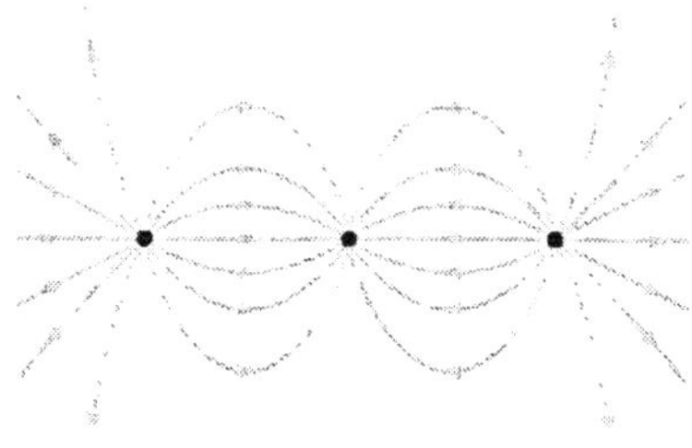

FIGURE 24-39 Problem 1 Solution.

Problem

3. Two charges $+q$ and a charge $-q$ are at the vertices of an equilateral triangle. Sketch some field lines for this charge distribution.

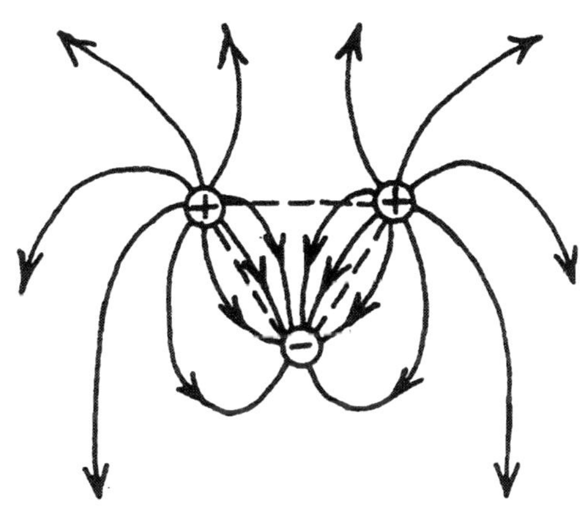

Problem 3 Solution.

Solution

(The sketch shown follows the text's convention of eight lines of force per charge magnitude q.)

Section 24-2: Electric Flux

Problem

5. A flat surface with area 2.0 m^2 is in a uniform electric field of 850 N/C. What is the electric flux through the surface when it is (a) at right angles to the field, (b) at 45° to the field, and (c) parallel to the field?

Solution

(a) When the surface is perpendicular to the field, its normal is either parallel or anti-parallel to **E**. Then Equation 24-1 gives $\Phi = \mathbf{E}\cdot\mathbf{A} = EA\cos(0° \text{ or } 180°) = \pm(850\ \text{N/C})(2\ \text{m}^2) = \pm1.70\ \text{kN}\cdot\text{m}^2/\text{C}$. (b) $\Phi = \mathbf{E}\cdot\mathbf{A} = EA\cos(45° \text{ or } 135°) = \pm(1.70\ \text{kN}\cdot\text{m}^2/\text{C})\times(0.866) = \pm1.20\ \text{kN}\cdot\text{m}^2/\text{C}$. (c) $\Phi = EA\cos 90° = 0$.

Problem

7. A flat surface with area 0.14 m^2 lies in the x-y plane, in a uniform electric field given by $\mathbf{E} = 5.1\hat{\mathbf{\imath}} + 2.1\hat{\mathbf{\jmath}} + 3.5\hat{\mathbf{k}}$ kN/C. Find the flux through this surface.

Solution

The surface can be represented by a vector area $\mathbf{A} = (0.14\ \text{m}^2)(\pm\hat{\mathbf{k}})$. (Since the surface is open, we have a choice of normal to the x-y plane.) Then $\Phi = \mathbf{E}\cdot\mathbf{A} = \pm\mathbf{E}\cdot\hat{\mathbf{k}}\times(0.14\ \text{m}^2) = \pm E_z(0.14\ \text{m}^2) = \pm(3.5\ \text{kN/C})(0.14\ \text{m}^2) = \pm490\ \text{N}\cdot\text{m}^2/\text{C}$. (Only the z component of the field contributes to the flux through the x-y plane.)

Problem

9. What is the flux through the hemispherical open surface of radius R shown in Fig. 24-41? The uniform field has magnitude E. *Hint:* Don't do a messy integral! Imagine closing the surface with a flat, circular piece across the open end. What would be the flux through the entire closed surface? And what's the flux through the flat end? So what's the answer?

Solution

All of the lines of force going through the hemisphere also go through an equitorial disk covering its edge in Fig. 24-41. Therefore, the flux through the disk (normal in the direction of **E**) equals the flux through the hemisphere. Since **E** is uniform, the flux through the disk is just $\pi R^2 E$. (Note: Gauss's Law gives the same result, since the flux through the closed surface, consisting of the hemisphere plus the disk, is zero. See Section 24-3.)

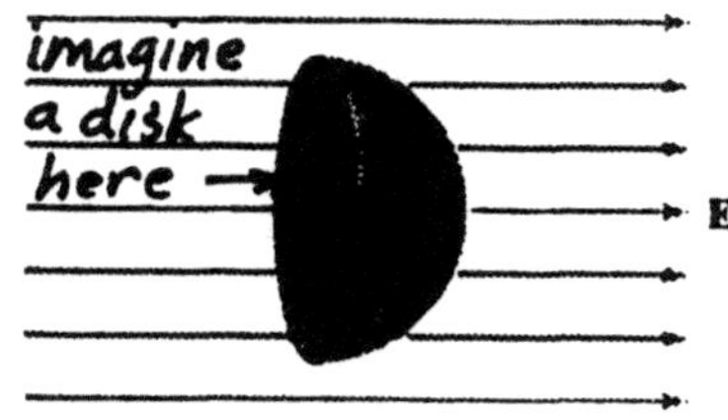

FIGURE 24-41 Problem 9 Solution.

Section 24-3: Gauss's Law

Problem

11. What is the electric flux through each closed surface shown in Fig. 24-43?

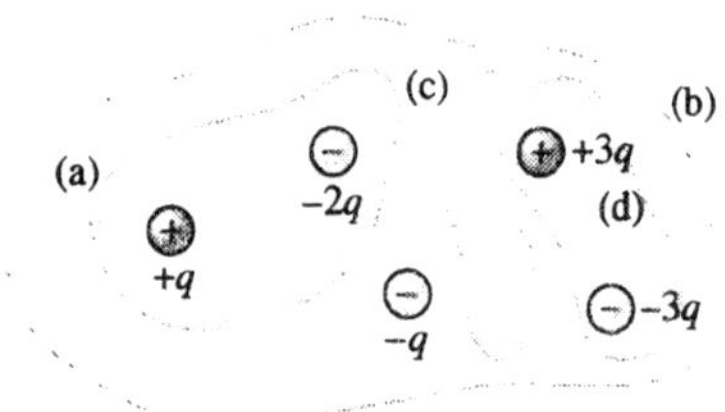

FIGURE 24-43 Problem 11.

Solution

From Gauss's law, $\Phi = q_{\text{enclosed}}/\varepsilon_0$. For the surfaces shown, this is (a) $(q - 2q)/\varepsilon_0 = -q/\varepsilon_0$, (b) $-2q/\varepsilon_0$, (c) and (d) 0.

Problem

13. A 2.6-μC charge is at the center of a cube 7.5 cm on each side. What is the electric flux through one face of the cube? *Hint:* Think about symmetry, and don't do an integral.

Solution

The symmetry of the situation guarantees that the flux through one face is $\frac{1}{6}$ the flux through the whole cubical surface, so $\Phi_{\text{face}} = \frac{1}{6}\oint_{\text{cube}} \mathbf{E}\cdot d\mathbf{A} = q_{\text{enclosed}}/6\varepsilon_0 = (2.6\ \mu\text{C})/6(8.85\times10^{-12}\ \text{C}^2/\text{N}\cdot\text{m}^2) = 49.0\ \text{kN}\cdot\text{m}^2/\text{C}$.

Problem

15. A dipole consists of two charges $\pm 6.1\ \mu$C located 1.2 cm apart. What is the electric flux through each surface shown in Fig. 24-44?

FIGURE 24-44 Problem 15.

Solution

It follows from Gauss's law that (a) $\Phi_a = +q/\varepsilon_0 = (6.1\ \mu\text{C})/(8.85\times10^{-12}\ \text{C}^2/\text{N}\cdot\text{m}^2) = 689\ \text{kN}\cdot\text{m}^2/\text{C}$, (b) $\Phi_b = -\Phi_a$, and (c) $\Phi_c = 0$.

Section 24-4: Using Gauss's Law

Problem

17. The electric field at the surface of a uniformly charged sphere of radius 5.0 cm is 90 kN/C. What would be the field strength 10 cm from the surface?

Solution

The electric field due to a uniformly charged sphere is like the field of a point charge for points outside the sphere, i.e., $E(r) \sim 1/r^2$ for $r \geq R$. Thus, at 10 cm from the surface, $r = 15$ cm and $E(15\ \text{cm}) = (5/15)^2 E\times(5\ \text{cm}) = (90\ \text{kN/C})/9 = 10\ \text{kN/C}$.

Problem

19. A crude model for the hydrogen atom treats it as a point charge $+e$ (the proton) surrounded by a uniform cloud of negative charge with total charge $-e$ and radius 0.0529 nm. What would be the electric field strength inside such an atom, halfway from the proton to the edge of the charge cloud?

Solution

At half the radius of the electron cloud, the field strength due to the cloud (a uniformly charged spherical volume) is given in Example 24-1: $E_e = k(-e)(\frac{1}{2}R)/R^3 = -ke/2R^2$. The field strength due to the proton (a point charge) at the same distance is $E_p = ke/(\frac{1}{2}R)^2 = 4ke/R^2$. (The electron's field is radially inward, negative, and the proton's is radially outward, positive.) The total field strength is $E = E_e + E_p = -ke/2R^2 + 4ke/R^2 = 7ke/2R^2 = 7(1.44\times10^{-9}\ \text{N}\cdot\text{m}^2/\text{C})/2(5.29\times10^{-11}\ \text{m})^2 = 1.80\times10^{12}\ \text{N/C}$.

Problem

21. A 10-nC point charge is located at the center of a thin spherical shell of radius 8.0 cm carrying −20 nC distributed uniformly over its surface. What are the magnitude and direction of the electric field (a) 2.0 cm, (b) 6.0 cm, and (c) 15 cm from the point charge?

Solution

The total electric field, the superposition of the fields due to the point charge and the spherical shell, is spherically symmetric about the center. Inside the shell ($r < R = 8$ cm), its field is zero, so the total field is just due to the 10 μC point charge. Outside ($r > R$), the shell's field is like that of a point charge of -20 μC at the same central location as the 10 μC charge. (This situation is described in Example 24-3.) (a) and (b) For $r = 2$ cm or 6 cm $< R$, $E = kq_{\text{pt.}}/r^2 = (9\times10^9\text{ N}\cdot\text{m}^2/C^2)(10\text{ nC})/(2\text{ cm or }6\text{ cm})^2 = 225\text{ kN/C}$ or 25.0 kN/C, respectively, directed radially outward. (c) For $r = 15\text{ cm} > R$, $E = k(q_{\text{pt}} + q_{\text{shell}})/r^2 = (9\times 10^9\text{ N}\cdot\text{m}^2/C^2)(10\text{ nC}-20\text{ nC})/(15\text{ cm})^2 = -4.00\text{ kN/C}$, directed radially inward.

Problem

23. A point charge $-2Q$ is at the center of a spherical shell of radius R carrying charge Q spread uniformly over its surface. What is the electric field at (a) $r = \frac{1}{2}R$ and (b) $r = 2R$? (c) How would your answers change if the charge on the shell were doubled?

Solution

The situation is like that in Problem 21. (a) At $r = \frac{1}{2}R < R$ (inside shell), $E = E_{\text{pt}} + E_{\text{shell}} = k(-2Q)/(\frac{1}{2}R)^2 + 0 = -8\,kQ/R^2$ (the minus sign means the direction is radially inward). (b) At $r = 2R > R$ (outside shell), $E = E_{\text{pt}} + E_{\text{shell}} = k(-2Q + Q)/(2R)^2 = -kQ/4R^2$ (also radially inward). (c) If $Q_{\text{shell}} = 2Q$, the field inside would be unchanged, but the field outside would be zero (since $q_{\text{shell}} + q_{\text{pt}} = 2Q - 2Q = 0$).

Problem

25. A spherical shell 30 cm in diameter carries a total charge 85 μC distributed uniformly over its surface. A 1.0-μC point charge is located at the center of the shell. What is the electric field strength (a) 5.0 cm from the center and (b) 45 cm from the center? (c) How would your answers change if the charge on the shell were doubled?

Solution

(a) The field due to the shell is zero inside, so at $r = 5$ cm, the field is due to the point charge only. Thus, $\mathbf{E} = kq\hat{\mathbf{r}}/r^2 = (9\times10^9\text{ N}\cdot\text{m}^2/\text{C}^2)(1\ \mu\text{C})\hat{\mathbf{r}}/(0.05\text{ m})^2 = (3.60\times10^6\text{ N/C})\hat{\mathbf{r}}$. (b) Outside the shell, its field is like that of a point charge, so at $r = 45$ cm, $\mathbf{E} = k(q + Q)\hat{\mathbf{r}}/r^2 = (9\times10^9\text{ N}\cdot\text{m}^2/\text{C}^2)(86\ \mu\text{C})\hat{\mathbf{r}}/(0.45\text{ m})^2 = (3.82\times10^6\text{ N/C})\hat{\mathbf{r}}$. (c) If the charge on the shell were doubled, the field inside would be unaffected, while the field outside would approximately double, $E = k(1.0\ \mu\text{C} + 2\times85\ \mu\text{C})/(45\text{ cm})^2 = 7.60\text{ MN/C}$.

Problem

27. How should the charge density within a solid sphere vary with distance from the center in order that the magnitude of the electric field in the sphere be constant?

Solution

Assume that ρ is spherically symmetric, and divide the volume into thin shells with $dV = 4\pi r^2\,dr$. From Gauss's law and Equation 24-5,

$$E = \frac{1}{4\pi\varepsilon_0 r^2}\int_V \rho\,dV = \frac{1}{4\pi\varepsilon_0 r^2}\int_0^r \rho(r')4\pi r'^2\,dr'$$
$$= \frac{1}{\varepsilon_0 r^2}\int_0^r \rho r'^2\,dr'.$$

It can be seen that if $\rho(r') \sim 1/r'$ then E is constant, but we can obtain the same result mathematically, by differentiation. If E is constant, $dE/dr = 0$. This implies

$$0 = \frac{d}{dr}\left(\frac{1}{r^2}\int_0^r \rho r'^2 dr'\right)$$
$$= \frac{1}{r^2}\frac{d}{dr}\left(\int_0^r \rho r'^2\,dr'\right) + \left(\int_0^r \rho r'^2\,dr'\right)\frac{d}{dr}\left(\frac{1}{r^2}\right)$$
$$= \frac{1}{r^2}\rho(r)r^2 - \frac{2}{r^3}\int_0^r \rho r'^2\,dr',$$

$$\text{or } \rho(r) = \frac{2}{r^3}\int_0^r \rho(r')r'^2\,dr'.$$

Since $r^{-2}\int_0^r \rho(r')r'^2\,dr' = \varepsilon_0 E$ is a constant, by hypothesis, $\rho(r) = 2\varepsilon_0 E/r \sim 1/r$, as suspected. (Look up how to take the derivative of an integral in any calculus textbook.) Note that constant magnitude does not imply constant direction; $\mathbf{E} = E\hat{\mathbf{r}}$ is spherically symmetric, not uniform.

Problem

29. A long solid rod 4.5 cm in radius carries a uniform volume charge density. If the electric field strength at the surface of the rod (not near either end) is 16 kN/C, what is the volume charge density?

Solution

If the rod is long enough to approximate its field using line symmetry, we can equate the flux through a length ℓ of its surface (Equation 24-8) to the charge enclosed. The latter is the charge density (a constant) times the volume of a length ℓ of rod. Thus, $2\pi R\ell E = q_{\text{enclosed}}/\varepsilon_0 = \rho\pi R^2\ell/\varepsilon_0$, or $\rho = 2\varepsilon_0 E/R = 2(8.85\times 10^{-12}\ \text{C}^2/\text{N}\cdot\text{m}^2)(16\ \text{kN/C})/(4.5\ \text{cm}) = 6.29\ \mu\text{C/m}^3$. (This is the magnitude of ρ, since the direction of the field at the surface, radially inward or outward, was not specified.)

Problem

31. An infinitely long rod of radius R carries a uniform volume charge density ρ. Show that the electric field strengths outside and inside the rod are given, respectively, by $E = \rho R^2/2\varepsilon_0 r$ and $E = \rho r/2\varepsilon_0$, where r is the distance from the rod axis.

Solution

The charge distribution has line symmetry (as in Problem 29) so the flux through a coaxial cylindrical surface of radius r (Equation 24-8) equals $q_{\text{enclosed}}/\varepsilon_0$, from Gauss's law. For $r > R$ (outside the rod), $q_{\text{enclosed}} = \rho\pi R^2\ell$, hence $E_{\text{out}} = \rho\pi R^2\ell/2\pi r\ell\varepsilon_0 = \rho R^2/2\varepsilon_0 r$. For $r < R$ (inside the rod), $q_{\text{enclosed}} = \rho\pi r^2\ell$, hence $E_{\text{in}} = \rho\pi r^2\ell/2\pi r\ell\varepsilon_0 = \rho r/2\varepsilon_0$. (The field direction is radially away from the symmetry axis if $\rho > 0$, and radially inward if $\rho < 0$.)

Problem

33. A long, thin wire carries a uniform line charge density $\lambda = -6.8\ \mu\text{C/m}$. It is surrounded by a thick concentric cylindrical shell of inner radius 2.5 cm and outer radius 3.5 cm. What uniform volume charge density in the shell will result in zero electric field outside the shell?

Solution

In order to have $\mathbf{E} = 0$ outside the shell, it is only necessary for the charge per unit length of shell to cancel that of the wire, i.e., $\lambda_{\text{shell}} = +6.8\ \mu\text{C/m}$ (see Gauss's law and Equation 24-8, with $q_{\text{enclosed}} = 0$). A uniform charge density which guarantees this is $\rho = \lambda_{\text{shell}}\ell/V$, where V is the volume of length ℓ of shell. Thus, $\rho = \lambda_{\text{shell}}\ell/\pi(r_2^2 - r_1^2)\ell = (6.8\ \mu\text{C/m})/\pi(3.5^2 - 2.5^2)\times 10^{-4}\ \text{m}^2 = 3.61\times 10^{-3}\ \text{C/m}^3$.

Problem

35. If you "painted" positive charge on the floor, what surface charge density would be necessary in order to suspend a 15-μC, 5.0-g particle above the floor?

Solution

A positive surface charge density σ, on the floor, would produce an approximately uniform electric field upward of $E = \sigma/2\varepsilon_0$, at points near the floor and not near an edge. The field needed to balance the weight of a particle, of mass m and charge q, is given by $mg = qE$, therefore $\sigma = 2\varepsilon_0 E = 2\varepsilon_0 mg/q = 2(8.85\times 10^{-12}\ \text{C}^2/\text{N}\cdot\text{m}^2)(5\times 10^{-3}\ \text{kg})(9.8\ \text{m/s}^2)/(15\times 10^{-6}\ \text{C}) = 57.8\ \text{nC/m}^2$.

Problem

37. Figure 24-47 shows sections of three infinite flat sheets of charge, each carrying surface charge density with the same magnitude σ. Find the magnitude and direction of the electric field in each of the four regions shown.

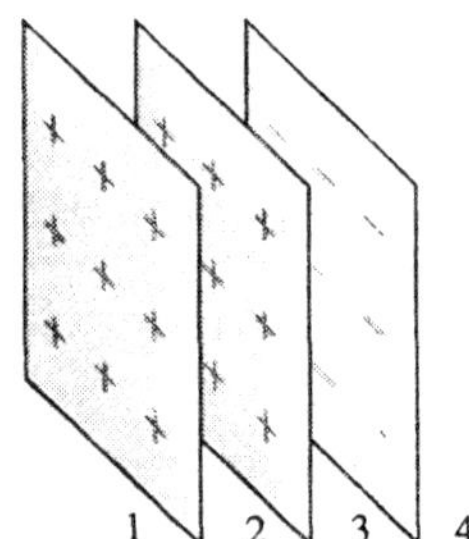

FIGURE 24-47 Problem 37.

Solution

The field from each sheet has magnitude $\sigma/2\varepsilon_0$ and points away from the positive sheets and toward the negative sheet. Take the x axis perpendicular to the sheets, to the right in Fig. 24-47. Superposition gives the field in each of the four regions, as shown.

	Region 1		Region 2		Region 3		Region 4
First sheet:	$=\xleftarrow{-\sigma\hat{\imath}/2\varepsilon_0}$	(+) (+)	$\xrightarrow{\sigma\hat{\imath}/2\varepsilon_0}$	(+) (+)	$\xrightarrow{\sigma\hat{\imath}/2\varepsilon_0}$	(−) (−)	$\xrightarrow{\sigma\hat{\imath}/2\varepsilon_0}$
Second sheet:	$\xleftarrow{-\sigma\hat{\imath}/2\varepsilon_0}$	(+) (+)	$\xleftarrow{-\sigma\hat{\imath}/2\varepsilon_0}$	(+) (+)	$\xrightarrow{\sigma\hat{\imath}/2\varepsilon_0}$	(−) (−)	$\xrightarrow{\sigma\hat{\imath}/2\varepsilon_0}$
Third sheet:	$\xrightarrow{-\sigma\hat{\imath}/2\varepsilon_0}$	(+) (+)	$\xrightarrow{\sigma\hat{\imath}/2\varepsilon_0}$	(+) (+)	$\xrightarrow{\sigma\hat{\imath}/2\varepsilon_0}$	(−) (−)	$\xleftarrow{-\sigma\hat{\imath}/2\varepsilon_0}$
Sum:	$\xleftarrow{-\sigma\hat{\imath}/2\varepsilon_0}$	(+) (+)	$\xrightarrow{\sigma\hat{\imath}/2\varepsilon_0}$	(+) (+)	$\xrightarrow{3\sigma\hat{\imath}/2\varepsilon_0}$	(−) (−)	$\xrightarrow{\sigma\hat{\imath}/2\varepsilon_0}$

Section 24-5: Fields of Arbitrary Charge Distributions

Problem

39. A nonconducting square plate 75 cm on a side carries a uniform surface charge density. The electric field strength 1 cm from the plate, not near an edge, is 45 kN/C. What is the approximate field strength 15 m from the plate?

Solution

The electric field strength close to the plate (1 cm $\ll$ 75 cm) has approximate plane symmetry ($E = \sigma/2\varepsilon_0$), so the charge on the plate is $q = \sigma A = 2\varepsilon_0 EA = 2(8.85\times10^{-12}\ \text{C}^2/\text{N}\cdot\text{m}^2)(45\ \text{kN/C})\times(0.75\ \text{m})^2 = 448\ \text{nC}$. Very far from the plate (15 m $\gg$ 0.75 m), the field strength is like that from a point charge, $E = kq/r^2 = (9\times10^9\ \text{N}\cdot\text{m}^2/\text{C}^2)\times(448\ \text{nC})(15\ \text{m})^{-2} = 17.9\ \text{N/C}$.

Problem

41. The electric field strength on the axis of a uniformly charged disk is given by $E = 2\pi k\sigma(1 - x/\sqrt{x^2 + a^2})$, with σ the surface charge density, a the disk radius, and x the distance from the disk center. If $a = 20$ cm, (a) for what range of x values does treating the disk as an infinite sheet give an approximation to the field that is good to within 10%? (b) For what range of x values is the point-charge approximation good to 10%?

Solution

(Note: The expression given, for the field strength on the axis of a uniformly charged disk, holds only for positive values of x.) (a) For small x, using the field strength of an infinite sheet, $E_{\text{sheet}} = \sigma/2\varepsilon_0 = 2\pi k\sigma$, produces a fractional error less than 10% if $|E_{\text{sheet}} - E|/E < 0.1$. Since $E_{\text{sheet}} > E$, this implies that $E_{\text{sheet}}/E < 1.1$ or $2\pi k\sigma/2\pi k\sigma(1 - x/\sqrt{x^2 + a^2}) < 1.1$. The steps in the solution of this inequality are: $1.1x < 0.1\sqrt{x^2 + a^2}$, $1.21x^2 < 0.01(x^2 + a^2)$, $x < a\sqrt{0.01/1.20} = 9.13\times10^{-2}a$. For $a = 20\ \text{cm}, x < 1.83$ cm. (b) For large x, the point charge field, $E_{\text{pt}} = kq/x^2 = k\pi\sigma a^2/x^2$, is good to 10% for $|E_{\text{pt}} - E|/E < 0.1$. The solution of this inequality is simplified by defining an angle ϕ, such that $\cos\phi = x/\sqrt{x^2 + a^2}$ and $\tan\phi = a/x$. In terms of ϕ, one finds $E = 2\pi k\sigma(1 - \cos\phi)$, $E_{\text{pt}} = k\pi\sigma\tan^2\phi$, and $E_{\text{pt}}/E = \tan^2\phi/2(1 - \cos\phi)$. Furthermore, $\tan^2\phi = \sin^2\phi/\cos^2\phi = (1 - \cos\phi)(1 + \cos\phi)/\cos^2\phi$, so $E_{\text{pt}}/E = (1 + \cos\phi)/2\cos^2\phi$. The range $0 \le x < \infty$ corresponds to $0 < \phi \le \pi/2$, so $E_{\text{pt}}/E > 1$ and the inequality becomes $E_{\text{pt}}/E = (1 + \cos\phi)/2\cos^2\phi < 1.1$, or $2.2\cos^2\phi - \cos\phi - 1 > 0$. The quadratic formula for the positive root gives $\cos\phi > (1 + \sqrt{1 + 8.8})/4.4 = 0.939$, or $\phi < 20.2°$. This implies $x = a/\tan\phi > a/\tan 20.2° = 2.72\ a$. For $a = 20$ cm, $x > 54.5$ cm.

Section 24-6: Gauss's Law and Conductors

Problem

43. What is the electric field strength just outside the surface of a conducting sphere carrying surface charge density 1.4 $\mu\text{C/m}^2$?

Solution

At the surface of a conductor, $E = \sigma/\varepsilon_0$ (positive away from the surface), or $(1.4\ \mu\text{C/m}^2)(8.85\times10^{-12}\ \text{C}^2/\text{N}\cdot\text{m}^2)^{-1} = 158$ kN/C in this problem.

Problem

45. A net charge of 5.0 μC is applied on one side of a solid metal sphere 2.0 cm in diameter. After electrostatic equilibrium is reached, what are (a) the volume charge density inside the sphere and (b) the surface charge density on the sphere? Assume there are no other charges or conductors nearby. (c) Which of your answers depends on this assumption, and why?

Solution

(a) The electric field within a conducting medium, in electrostatic equilibrium, is zero. Therefore, Gauss's law implies that the net charge contained in any closed surface, lying within the metal, is zero. (b) If the volume charge density is zero within the metal, all of the net charge must reside on the surface of the sphere. If the sphere is electrically isolated, the charge will be uniformly distributed (i.e., spherically symmetric), so $\sigma = Q/4\pi R^2 = (5\ \mu\text{C})/4\pi(1\ \text{cm})^2 = 3.98\times10^{-3}\ \text{C/m}^2$. (c) Spherical symmetry for σ depends on the proximity of other charges and conductors.

Problem

47. A 250 nC point charge is placed at the center of an uncharged spherical conducting shell 20 cm in radius. (a) What is the surface charge density on the outer surface of the shell? (b) What is the electric field strength at the shell's outer surface?

Solution

(a) There is a non-zero field outside the shell, because the net charge within is not zero. Therefore, there is a surface charge density $\sigma = \varepsilon_0 E$ on the outer surface of the shell, which is uniform, if we ignore the possible presence of other charges and conducting surfaces outside the shell. Gauss's law (with reasoning similar to Example 24-7) requires that the charge on the shell's outer surface is equal to the point charge within, so $\sigma = q/4\pi R^2 = 250\ \text{nC}/4\pi(0.20\ \text{m})^2 = 497\ \text{nC/m}^2$. (b) Then the field strength at the outer surface is $E = \sigma/\varepsilon_0 = 56.2$ kN/C.

Problem

49. An irregular conductor containing an irregular, empty cavity carries a net charge Q. (a) Show that the electric field inside the cavity must be zero.

(b) If you put a point charge inside the cavity, what value must it have in order to make the surface charge density on the outer surface of the conductor everywhere zero?

Solution

(a) When there is no charge inside the cavity, the flux through any closed surface within the cavity (S_1) is zero, hence so is the field. (b) If the surface charge density on the outer surface (and also the electric field there) is to vanish, then the net charge inside a gaussian surface containing the conductor (S_2) is zero. Thus, the point charge in the cavity must equal $-Q$. (Note: The argument in part (a) depends on the conservative nature of the electrostatic field (see Section 25-1), for then positive flux on one part of S_1 canceling negative flux on another part is ruled out.)

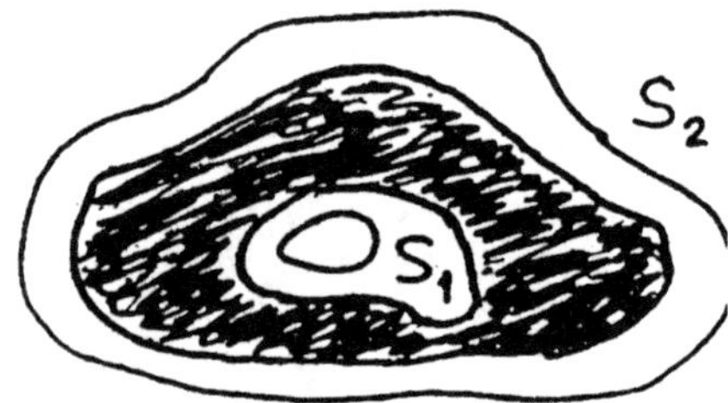

Problem 49 Solution.

Problem

51. A total charge of 18 μC is applied to a thin, square metal plate 75 cm on a side. Find the electric field strength near the plate's surface.

Solution

The net charge of 18 μC must distribute itself over the outer surface of the plate, in accordance with Gauss's law for conductors. The outer surface consists of two plane square surfaces on each face, plus the edges and corners. Symmetry arguments imply that for an isolated plate, the charge density on the faces is the same, but not necessarily uniform because the edges and corners also have charge. If the plate is thin, we could assume that the edges and corners have negligible charge and that the density on the faces is approximately uniform. Then the surface charge density is the total charge divided by the area of both faces, $\sigma = 18\ \mu\text{C}/2(75\text{ cm})^2 = 16.0\ \mu\text{C}$, and the field strength near the plate (but not near an edge) is $E = \sigma/\varepsilon_0 = 1.81\text{ MN/C}$.

Problem

53. A conducting sphere 2.0 cm in radius is concentric with a spherical conducting shell with inner radius 8.0 cm and outer radius 10 cm. The small sphere carries 50 nC charge and the shell has no net charge. Find the electric field strength (a) 1.0 cm, (b) 5.0 cm, (c) 9.0 cm, and (d) 15 cm from the center.

Solution

If we assume the two-conductor system is isolated and in electrostatic equilibrium, then the field has spherical symmetry. Gauss's law requires that the field inside the conducting material be zero (for $0 \leq r < 2$ cm and 8 cm $< r <$ 10 cm in this problem), and that, since the shell is neutral, the field elsewhere is like that from a point charge of 50 μC located at the center of symmetry ($r = 0$). Thus, (a) $E(1\text{ cm}) = 0$, (b) $E(5\text{ cm}) = kq/r^2 = (9\times10^9\text{ N}\cdot\text{m}^2/\text{C}^2)\times$ $(50\ \mu\text{C})/(5\text{ cm})^2 = 180\text{ kN/C}$, (c) $E(9\text{ cm}) = 0$, and (d) $E(15\text{ cm}) = kq/r^2 = (\frac{1}{9})E(5\text{ cm}) = 20\text{ kN/C}$.

Paired Problems

Problem

55. A point charge $-q$ is at the center of a spherical shell carrying charge $+2q$. That shell, in turn, is concentric with a larger shell carrying charge $-\frac{3}{2}q$. Draw a cross section of this structure, and sketch the electric field lines using the convention that 8 lines correspond to a charge of magnitude q.

Solution

The field from the given charges is spherically symmetric, so (from Gauss's law) is like that of a point charge, located at the center, with magnitude equal to the net charge enclosed by a sphere of radius

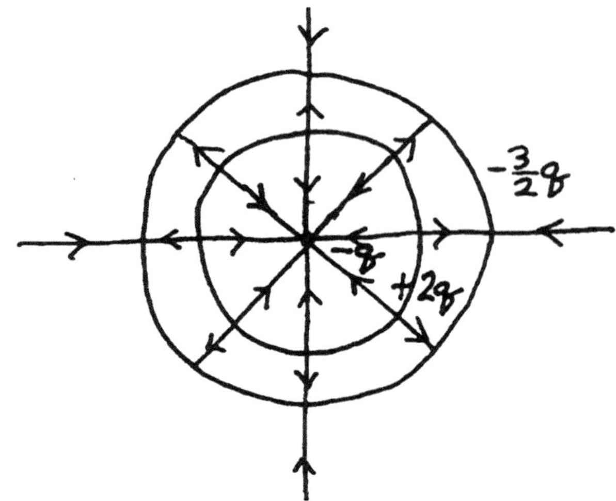

Problem 55 Solution.

equal to the distance to the field point. Thus, $E = -kq/r^2$ inside the first shell (8 lines radially inward), $E = +kq/r^2$ between the first and second shells (8 lines radially outward), and $E = -kq/2r^2$ outside the second shell (4 lines radially inward).

Problem

57. A point charge q is at the center of a spherical shell of radius R carrying charge $2q$ spread uniformly over its surface. Write expressions for the electric field strength at (a) $\frac{1}{2}R$ and (b) $2R$.

Solution

As explained in Example 24-3, (a) for $r = \frac{1}{2}R < R$, $q_{\text{enclosed}} = q$ and $E = kq/(\frac{1}{2}R)^2 = 4kq/R^2$, and (b) for $r = 2R > R$, $q_{\text{enclosed}} = q + 2q$ and $E = 3kq/(2R)^2 = 3\ kq/4R^2$.

Problem

59. A long, thin hollow pipe 4.0 cm in diameter carries charge at a density of $-2.6\ \mu$C/m, uniformly distributed over the pipe. It is concentric with 10-cm diameter pipe carrying $+2.6\ \mu$C/m, also uniformly distributed. Find the magnitude of the electric field at (a) 0.50 cm, (b) 3.5 cm, and (c) 12 cm from the axis of the pipes.

Solution

Assume the electric field has line symmetry, and apply Gauss's law to a coaxial cylindrical surface of radius r. The result is Equation 24-8 set equal to $\lambda_{\text{enclosed}}\ell/\varepsilon_0$, so $E(r) = \lambda_{\text{enclosed}}/2\pi\varepsilon_0 r$. (a) At $r = 0.5$ cm < 2 cm (inside inner pipe), $\lambda_{\text{enclosed}}$ is zero and so is E. (b) At 2 cm $< r = 3.5$ cm < 5 cm (between the pipes), $\lambda_{\text{enclosed}}$ is just the inner pipe, so $E = (-2.6\ \mu\text{C/m}) \div 2\pi\varepsilon_0(3.5\text{ cm}) = -1.34$ MN/C. (The minus sign means the direction of E is radially inward toward the axis of the pipes; the magnitude is the absolute value of E.) (c) At $r = 12$ cm > 5 cm (outside outer pipe), $\lambda_{\text{enclosed}}$ and E are again zero, since the pipes have opposite linear charge densities.

Problem

61. An early (and incorrect) model for the atom pictured its positive charge as spread uniformly throughout the spherical atomic volume. For a hydrogen atom of radius 0.0529 nm, what would be the electric field due to such a distribution of positive charge (a) 0.020 nm from the center and (b) 0.20 nm from the center?

Solution

(a) Inside a uniformly charged spherical volume, $E = kQr/R^3 = (9\text{ GN}\cdot\text{m}^2/\text{C}^2)(1.6\times10^{-19}\text{ C})\times(0.02\text{ nm})/(0.0529\text{ nm})^3 = 195$ GN/C (see Equation 24-7). (b) Outside, the field is like that of a point charge, $E = kQ/r^2 = (9\text{ GN}\cdot\text{m}^2/\text{C}^2)(1.6\times10^{-19}\text{ C}) \div (0.2\text{ nm})^2 = 36.0$ GN/C (see Equation 24-6).

Problem

63. A sphere of radius $2a$ has a hole of radius a, as shown in Fig. 24-50. The solid portion carries a uniform volume charge density ρ. Find an expression for the electric field strength within the solid portion, as a function of the distance r from the center.

Solution

From Gauss's law and Equation 24-5, $E = q_{\text{enclosed}} \div 4\pi\varepsilon_0 r^2$, where q_{enclosed} is the charge within a spherical gaussian surface of radius r about the center of symmetry. For $a < r < 2a$, $q_{\text{enclosed}} = \rho V = \frac{4}{3}\pi\rho(r^3 - a^3)$, so $E = (\rho/3\varepsilon_0)(r - a^3/r^2)$.

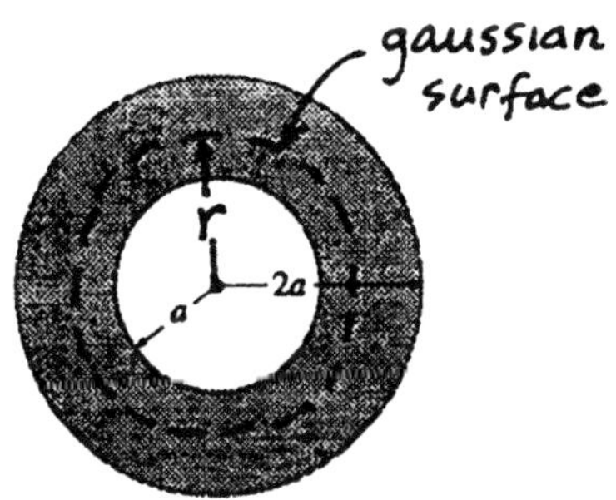

FIGURE 24-50 Problem 63 Solution.

Supplementary Problems

Problem

65. Repeat Problem 10 for the case $\mathbf{E} = E_0\left(\frac{y}{a}\right)^2\hat{\mathbf{k}}$.

Solution

Since the electric field depends only on y, break up the square in Fig. 24-42 (see Problem 10) into strips of area $d\mathbf{A} = \pm a\ dy\ \hat{\mathbf{k}}$, of length a parallel to the x axis and width dy, the normal to which could be $\pm\hat{\mathbf{k}}$. The electric flux through the square is

$$\Phi = \int_{\text{square}} \mathbf{E}\cdot d\mathbf{A} = \pm\int_0^a E_0\left(\frac{y}{a}\right)^2 a\ dy$$
$$= \pm\left(\frac{E_0}{a}\right)\int_0^a y^2\ dy = \pm\frac{1}{3}E_0 a^2.$$

Problem

67. A proton is released from rest 1.0 cm from a large sheet carrying a surface charge density of $-24\ \text{nC/m}^2$. How much later does it strike the sheet?

Solution

The proton is accelerated toward the sheet by an electric field in that direction. For the field, we can use that from an infinite plane sheet (assuming we are not near an edge) so $E = \sigma/2\varepsilon_0$ and the acceleration $a = eE/m_p$ is uniform. Starting from rest, the proton travels to the sheet in time

$$t = \sqrt{2(x - x_0)/a} = \sqrt{4\varepsilon_0 m_p(x - x_0)/e\sigma}$$

$$= \sqrt{\frac{4(8.85\times10^{-12}\ \text{C}^2/\text{N}\cdot\text{m}^2)(1.67\times10^{-27}\ \text{kg})(1\ \text{cm})}{(1.6\times10^{-19}\ \text{C})(24\ \text{nC/m}^2)}}$$

$= 392$ ns.

Problem

69. Repeat Problem 36 for the case when the charge density in the slab is given by $\rho = \rho_0\,|x/d|$, where ρ_0 is a constant.

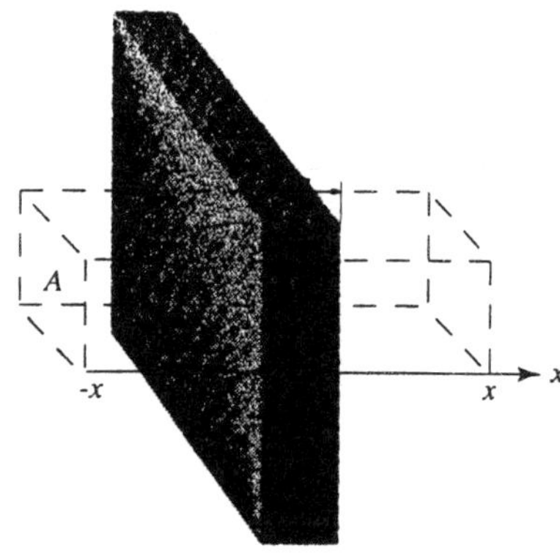

FIGURE 24-45 Problem 69 Solution.

Solution

Gauss's law, plane symmetry, and Equation 24-9 can be used to find the electric field strength, but we must integrate to get the charge enclosed by the gaussian surface. We use charge elements that are thin parallel sheets of the same area as the face of the gaussian surface and of thickness dx, as shown. (a) Inside the slab ($|x| < d/2$), $2EA = \varepsilon_0^{-1}\int_{-x}^{x} \rho A\,dx = (\rho_0 A/\varepsilon_0 d)\times(\int_{-x}^{0}(-x)\,dx + \int_0^x x\,dx) = \rho_0 Ax^2/\varepsilon_0 d$, hence $E = \rho_0 x^2/2\varepsilon_0 d$. (b) Outside, $2EA = \varepsilon_0^{-1}\int_{-d/2}^{d/2}\rho A\,dx = (\rho_0 A/\varepsilon_0 d)(\int_{-d/2}^{0}(-x)\,dx + \int_0^{d/2} x\,dx) = (\rho_0 A/\varepsilon_0 d)(d/2)^2$, hence $E = \rho_0 d/8\varepsilon_0$. (This is equivalent, of course, to the field strength outside an infinite sheet with $\sigma = \rho_0 d/4$.)

Problem

71. A small object of mass m and charge q is attached by a thread of length ℓ to a large, flat, nonconducting plate carrying a uniform surface charge density σ with the same sign as q (Fig. 24-52). If the object is displaced slightly sideways from its equilibrium, show that it undergoes simple harmonic motion with period $T = 2\pi\sqrt{2\varepsilon_0 m\ell/q\sigma}$. Assume the gravitational force is negligible.

FIGURE 24-52 Problem 71.

Solution

The Coulomb force on the particle is approximately uniform and greater than its weight. Therefore, the net upward force field on the particle is $F_{\text{net}}/m = (q/m)E - g$, and since it is tethered by the string, it oscillates like an upside-down pendulum with period $T = 2\pi\sqrt{\ell/(qE/m - g)}$. If we neglect gravity and use $E = \sigma/2\varepsilon_0$ for an infinite sheet as an approximation, then $T = 2\pi\sqrt{2\varepsilon_0 m\ell/q\sigma}$.

Problem

73. A thick spherical shell of inner radius a and outer radius b carries a charge density given by $\rho = \dfrac{ce^{-r/a}}{r^2}$, where a and c are constants. Find expressions for the electric field strength for (a) $r < a$, (b) $a < r < b$, and (c) $r > b$.

Solution

Spherical symmetry, Equation 24-5 and Gauss's law give a field strength of $E(r) = q_{\text{enclosed}}/4\pi\varepsilon_0 r^2$, where $q_{\text{enclosed}} = \int_0^r \rho\,dV$ is the charge within a concentric spherical surface of radius r, and $dV = 4\pi r^2\,dr$ is the volume element for a thin shell with this surface.
(a) For $r < a$, $q_{\text{enclosed}} = 0$ hence $E(r) = 0$. (b) For $a \le r \le b$, $q_{\text{enclosed}} = 4\pi c\int_a^r e^{-r/a}\,dr = 4\pi ac(e^{-1} - e^{r/a})$ hence $E(r) = ac(e^{-1} - e^{-r/a})/\varepsilon_0 r^2$. (c) For $r > b$, $q_{\text{enclosed}} = 4\pi c\int_a^b e^{-r/a}\,dr$ hence $E(r) = ac(e^{-1} - e^{-b/a})/\varepsilon_0 r^2$.

Problem

75. A solid sphere of radius R carries a uniform volume charge density ρ. A hole of radius $R/2$ occupies a region from the center to the edge of the sphere, as shown in Fig. 24-53. Show that the electric field everywhere in the hole points horizontally and has magnitude $\rho R/6\varepsilon_0$. *Hint:* Treat the hole as a superposition of two charged spheres of opposite charge.

Solution

A large solid sphere can be considered to be the superposition of the sphere with a cavity plus a small solid sphere filling the cavity, both with uniform charge density ρ. The electric field inside the solid spheres is $\rho\mathbf{r}/3\varepsilon_0$, where $\mathbf{r}$ is a vector from the center of each sphere to the field point P, in both (see Equation 24-7). For the large sphere, whose center we take at the origin, $\mathbf{r} = \mathbf{r}_P$, and for the small sphere, whose center is at $\frac{1}{2}R\hat{\imath}, \mathbf{r} = \mathbf{r}_P - \frac{1}{2}R\hat{\imath}$. Therefore, $\mathbf{E}$(large sphere) $=$ $\mathbf{E}$(sphere with cavity) $+$ $\mathbf{E}$(small sphere), or $\rho\mathbf{r}_P/3\varepsilon_0 = \mathbf{E} + \rho(\mathbf{r}_P - \frac{1}{2}R\hat{\imath})/3\varepsilon_0$. Thus, $\mathbf{E} = \rho R\hat{\imath}/6\varepsilon_0$, that is, for any point inside the cavity, the electric field of the sphere with the cavity is uniform (with direction parallel to the line between the centers of the sphere and cavity). (Note that this result holds for any size spherical cavity if one replaces $\frac{1}{2}R\hat{\imath}$ with the vector to the center of the cavity.)

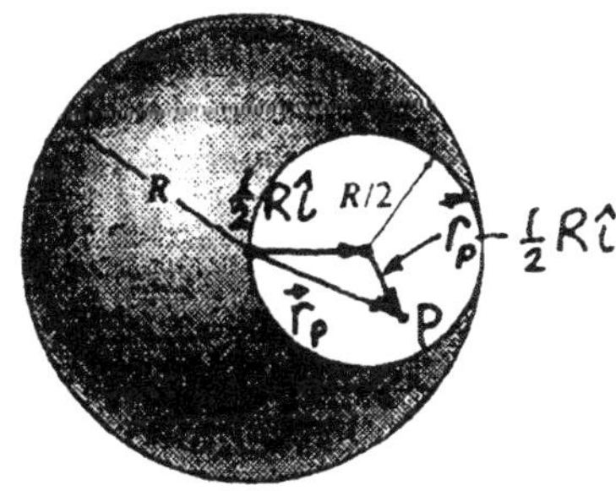

FIGURE 24-53 Problem 75 Solution.

Problem

77. Two flat, parallel, closely spaced metal plates of area 0.080 m^2 carry total charges of $-2.1\ \mu C$ and $+3.8\ \mu C$. Find the surface charge densities on the inner and outer faces of each plate.

Solution

If the thickness and separation of the plates is small compared to their lateral dimensions ($\sqrt{0.08\ m^2} \approx$ 28 cm), we can assume that the electric field near the plates (edge effects neglected) is uniform and normal to the plates. (Of course, far away, the field goes like $1/r^2$.) The general case, where the plates carry arbitrary charges q_1 and q_2, can be viewed as the superposition of three simpler cases: a neutral plate, a charged plate, and two oppositely charged plates (see last paragraph in Section 24-6). In each of the three regions, left of, between, and right of the plates (the thickness, assumed negligible, does not add to any region), the electric field is the sum of three contributions, as shown. The final diagram, together with Equation 24-11 (which gives $\sigma = \varepsilon_0 E$ at each surface), shows that the charge densities on the outer surfaces are equal, $\sigma_{out} = \frac{1}{2}(q_1 + q_2)/A$, and that the charge densities on the inner surfaces are equal and opposite, $\sigma_{in} = \pm\frac{1}{2}(q_2 - q_1)/A$. Numerically, $\sigma_{out} = \frac{1}{2}(-2.1 + 3.8)\ \mu C/0.08\ m^2 = 10.6\ \mu C/m^2$, and $\sigma_{in} = \pm\frac{1}{2}[3.8 - (-2.1)]\ \mu C/0.08\ m^2 = \pm 36.9\ \mu C/m^2$. (Note: the direction of the fields is shown for positive total charge, with the more positive plate on the right.)

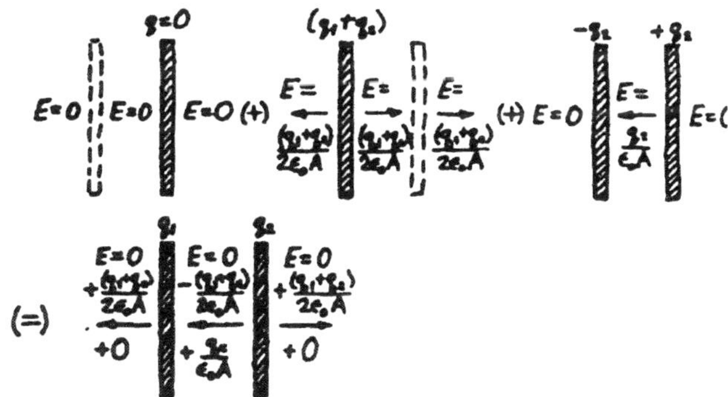

Problem 77 Solution.

CHAPTER 25 ELECTRIC POTENTIAL

ActivPhysics can help with these problems: Activities 11.9, 11.10

Section 25-2: Potential Difference

Problem

1. How much work does it take to move a 50-μC charge against a 12-V potential difference?

Solution

The potential difference and the work per unit charge, done by an external agent, are equal in magnitude, so $W = q\ \Delta V = (50\ \mu\text{C})(12\ \text{V}) = 600\ \mu\text{J}$. (Note: Since only magnitudes are needed in this problem, we omitted the subscripts A and B.)

Problem

3. It takes 45 J to move a 15-mC charge from point A to point B. What is the potential difference ΔV_{AB}?

Solution

The work done by an external agent equals the potential energy change, $\Delta U_{AB} = 45\ \text{J} = q\ \Delta V_{AB}$, hence $\Delta V_{AB} = 45\ \text{J}/15\ \text{mC} = 3\ \text{kV}$. (Since the work required to move the charge from A to B is positive, $V_B > V_A$ and ΔV_{AB} is positive.)

Problem

5. Find the magnitude of the potential difference between two points located 1.4 m apart in a uniform 650 N/C electric field, if a line between the points is parallel to the field.

Solution

For ℓ in the direction of a uniform electric field, Equation 25-2b gives $|\Delta V| = E\ell = (650\ \text{N/C})\times(1.4\ \text{m}) = 910\ \text{V}$. (See note in solution to Problem 1. Since $dV = -\mathbf{E}\cdot d\boldsymbol{\ell}$, the potential always decreases in the direction of the electric field.)

Problem

7. Two points A and B lie 15 cm apart in a uniform electric field, with the path AB parallel to the field. If the potential difference ΔV_{AB} is 840 V, what is the field strength?

Solution

Equation 25-2b for a uniform field gives $E = |\Delta V/\ell| = 840\ \text{V}/0.15\ \text{m} = 5.60\ \text{kV/m}$. (See notes in solutions to Problems 1 and 5.)

Problem

9. A proton, an alpha particle (a bare helium nucleus), and a singly ionized helium atom are accelerated through a potential difference of 100 V. Find the energy each gains.

Solution

The energy gained is $q\ \Delta V$ (see Example 25-1). The proton and singly-ionized helium atom have charge e, so they gain $100\ \text{eV} = (1.6\times10^{-19}\ \text{C})(100\ \text{V}) = 1.6\times10^{-17}\ \text{J}$, while the α-particle has charge $2e$ and gains twice this energy.

Problem

11. What is the potential difference between the terminals of a battery that can impart 7.2×10^{-19} J to each electron that moves between the terminals?

Solution

The magnitude of the potential difference is $|W/q| = 7.2\times10^{-19}\ \text{J}/1.6\times10^{-19}\ \text{C} = 4.5\ \text{V}$. (The energy imparted per electron is 4.5 eV.)

Problem

13. A 12-V car battery stores 2.8 MJ of energy. How much charge can move between the battery terminals before it is totally discharged? Assume the potential difference remains at 12 V, an assumption that is not realistic.

Solution

A charge q, moving through a potential difference ΔV, is equivalent to electrostatic potential energy $\Delta U = q\ \Delta V$, stored in the battery. Thus, $q = 2.8\ \text{MJ}/12\ \text{V} = 2.33\times10^5\ \text{C}$.

Problem

15. Two large, flat metal plates are a distance d apart, where d is small compared with the plate size. If

the plates carry surface charge densities $\pm\sigma$, show that the potential difference between them is $V = \sigma d/\varepsilon_0$.

Solution

The electric field between the plates is uniform, with $E = \sigma/\varepsilon_0$, directed from the positive to the negative plate (see last paragraph of Section 24-6 and Fig. 24-35). Then Equation 25-2b gives $V = V_+ - V_- = -(\sigma/\varepsilon_0)(-d) = \sigma d/\varepsilon_0$ (the displacement from the negative to the positive plate is opposite to the field direction).

Problem

17. A 5.0-g object carries a net charge of 3.8 μC. It acquires a speed v when accelerated from rest through a potential difference V. A 2.0-g object acquires twice the speed under the same circumstances. What is its charge?

Solution

The speed acquired by a charge q, starting from rest at point A and moving through a potential difference of V, is given by $\frac{1}{2}mv_B^2 = q(V_A - V_B) = qV$, or $v_B = \sqrt{2V(q/m)}$. (This is the work-energy theorem for the electric force. A positive charge is accelerated in the direction of decreasing potential.) If the second object acquires twice the speed of the first object, moving through the same potential difference, it must have four times the charge to mass ratio, q/m. Thus, $q_2 = 4(q_1/m_1)m_2 = 4(3.8\ \mu\text{C})(2\text{g}/5\text{g}) = 6.08\ \mu\text{C}$.

Section 25-3: Calculating Potential Difference

Problem

19. The classical picture of the hydrogen atom has a single electron in orbit a distance 0.0529 nm from the proton. Calculate the electric potential associated with the proton's electric field at this distance.

Solution

The potential of the proton, at the position of the electron (both of which may be regarded as point-charge atomic constituents) is (Equation 25-4) $V = ke/a_0$, where a_0 is the Bohr radius. Numerically, $V = (9\times10^9\ \text{N}\cdot\text{m}^2/\text{C}^2)(1.6\times10^{-19}\ \text{C})/(5.29\times10^{-11}\ \text{m}) = 27.2$ V. (The energy of an electron in a classical, circular orbit, around a stationary proton, is one half its potential energy, or $\frac{1}{2}U = \frac{1}{2}(-e)V = -13.6$ eV. The excellent agreement with the ionization energy of hydrogen was one of the successes of the Bohr model.)

Problem

21. Points A and B lie 20 cm apart on a line extending radially from a point charge Q, and the potentials at these points are $V_A = 280$ V, $V_B = 130$ V. Find Q and the distance r between A and the charge.

Solution

Since $V_A = kQ/r_A$ and $V_B = kQ/r_B$, division yields $r_B = (V_A/V_B)r_A = (280/130)r_A = 2.15r_A$. But $r_B - r_A = 20$ cm, so $r_A = (20\ \text{cm})(2.15 - 1)^{-1} = 17.3$ cm. Then $Q = V_A r_A/k = (280\ \text{V})(17.3\ \text{cm})\div(9\times10^9\ \text{N}\cdot\text{m}^2/\text{C}^2) = 5.39$ nC.

Problem

23. A 3.5-cm-diameter isolated metal sphere carries a net charge of 0.86 μC. (a) What is the potential at the sphere's surface? (b) If a proton were released from rest at the sphere's surface, what would be its speed far from the sphere?

Solution

(a) An isolated metal sphere has a uniform surface charge density, so Equation 25-4 gives the potential at its surface, $V_{\text{surf}} = kQ/R = (9\ \text{GN}\cdot\text{m}^2/\text{C}^2)(0.86\ \mu\text{C})\div(\frac{1}{2}\times3.5\ \text{cm}) = 442$ kV. (b) The work done by the repulsive electrostatic field (the negative of the change in the proton's potential energy) equals the proton's kinetic energy at infinity, $W = -e(V_\infty - V_{\text{surf}}) = eV_{\text{surf}} = \frac{1}{2}mv^2$. Then $v = \sqrt{2eV_{\text{surf}}/m} = [2(1.6\times10^{-19}\ \text{C})(442\ \text{kV})/(1.67\times10^{-27}\ \text{kg})]^{1/2} = 9.21$ Mm/s.

Problem

25. A thin spherical shell of charge has radius R and total charge Q distributed uniformly over its surface. What is the potential at its center?

Solution

The potential at the surface of the shell is kQ/R (as in Example 25-3). The electric field inside a uniformly charged shell is zero, so the potential anywhere inside is a constant, equal, therefore, to its value at the surface.

Problem

27. Find the potential as a function of position in an electric field given by $\mathbf{E} = ax\,\hat{\mathbf{i}}$, where a is a constant and where $V = 0$ at $x = 0$.

Solution

Since $V(0) = 0$, $V(\mathbf{r}) = -\int_0^r \mathbf{E}\cdot d\mathbf{r} = -\int_0^r ax\hat{\mathbf{i}}\cdot d\mathbf{r} = -\int_0^x ax\,dx = -\frac{1}{2}ax^2$.

Problem

29. The potential difference between the surface of a 3.0-cm-diameter power line and a point 1.0 m distant is 3.9 kV. What is the line charge density on the power line?

Solution

If we approximate the potential from the line by that from an infinitely long charged wire, Equation 25-5 can be used to find λ: $\lambda = 2\pi\varepsilon_0\,\Delta V_{AB}/\ln(r_A/r_B) = (3.9\text{ kV})[2(9\times10^9\text{ N}\cdot\text{m}^2/\text{C}^2)\ln(100/1.5)]^{-1} = 51.6\text{ nC/m}$. (Note: $\Delta V_{AB} = V_B - V_A$ so B is at the surface of the wire and A is 100 cm distant.)

Problem

31. A charge $+Q$ lies at the origin, and $-3Q$ at $x = a$. Find two points on the x-axis where $V = 0$.

Solution

The potential on the x-axis, $\Sigma\, kq_i/r_i = kQ/|x| + k(-3Q)/|x-a|$ (from Equation 25-6), is zero when $3|x| = |x-a|$. For $x < 0$, this implies $-3x = a - x$, or $x = -a/2$. Note that the absolute value of a negative number is minus the number, and we assume that a is a positive constant. For $0 < x < a$, the condition is $3x = a - x$, or $x = a/4$. For $x > a$, there are no solutions. (The same results follow from the quadratic $8x^2 + 2ax - a^2 = 0$, which results from the square of the above condition.)

Problem

33. Find the potential 10 cm from a dipole of moment $p = 2.9$ nC·m (a) on the dipole axis, (b) at 45° to the axis, and (c) on the perpendicular bisector. The dipole separation is much less than 10 cm.

Solution

Equation 25-7 gives the potential from a point dipole as a function of distance and angle from the dipole axis. For the dipole moment and distance given, $V(r,\theta) = kp\cos\theta/r^2 = (9\text{ GN}\cdot\text{m}^2/\text{C}^2)(2.9\text{ nC}\cdot\text{m})\times\cos\theta/(10\text{ cm})^2 = (2.61\text{ kV})\cos\theta$. For the three given angles,
(a) $V = (2.61\text{ kV})\cos 0° = 2.61$ kV;
(b) $V = (2.61\text{ kV})\cos 45° = 1.85$ kV; and
(c) $V = (2.61\text{ kV})\cos 90° = 0$.

Problem

35. A hollow, spherical conducting shell of inner radius b and outer radius c surrounds, and is concentric with, a solid conducting sphere of radius a, as shown in Fig. 25-39. The sphere carries a net charge $-Q$ and the shell carries a net charge $+3Q$. Both conductors are in electrostatic equilibrium. Find an expression for the potential difference from infinity to the surface of the sphere.

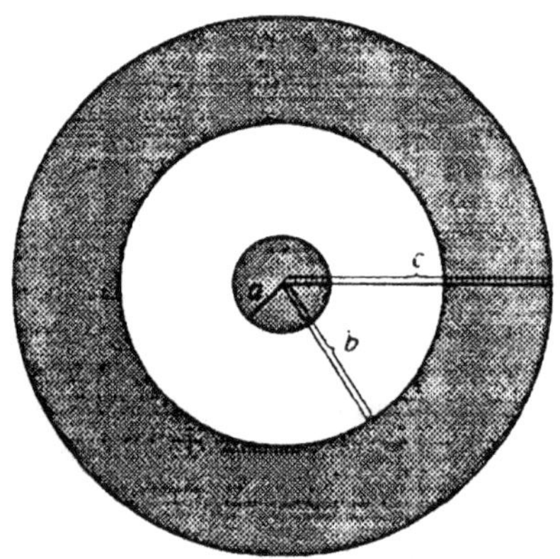

FIGURE 25-39 Problem 35.

Solution

The electric field between the solid sphere and the shell is like that due to a point charge $-Q$ located at their common center (the origin), so the potential difference between the sphere and the shell is $V_{\text{sph}} - V_{\text{shell}} = -kQ(a^{-1} - b^{-1})$. The electric field outside the shell is like that due to a point charge $2Q$ at the origin (the total charge is $3Q - Q$), so the potential difference between the shell and infinity is $V_{\text{shell}} - V_\infty = 2kQc^{-1}$. (See Examples 24-3 and 25-3.) The entire shell is at one potential, as is the entire sphere, because each is a conductor in electrostatic equilibrium. Therefore, the potential difference between the sphere and infinity is $V_{\text{sph}} - V_\infty = V_{\text{sph}} - V_{\text{shell}} + V_{\text{shell}} - V_\infty = kQ(2c^{-1} + b^{-1} - a^{-1})$.

Problem

37. A thin ring of radius R carries a charge $3Q$ distributed uniformly over three-fourths of its circumference, and $-Q$ over the rest. What is the potential at the center of the ring?

Solution

The result in Example 25-6 did not depend on the ring being uniformly charged. For a point on the axis of the ring, the geometrical factors are the same, and $\int_{\text{ring}} dq = Q_{\text{tot}}$ for any arbitrary charge distribution, so $V = kQ_{\text{tot}}(x^2 + a^2)^{-1/2}$ still holds. Thus, at the center ($x = 0$) of a ring of total charge $Q_{\text{tot}} = 3Q - Q = 2Q$, and radius $a = R$, the potential is $V = 2kQ/R$.

Problem

39. The annulus shown in Fig. 25-40 carries a uniform surface charge density σ. Find an expression for the potential at an arbitrary point P on its axis.

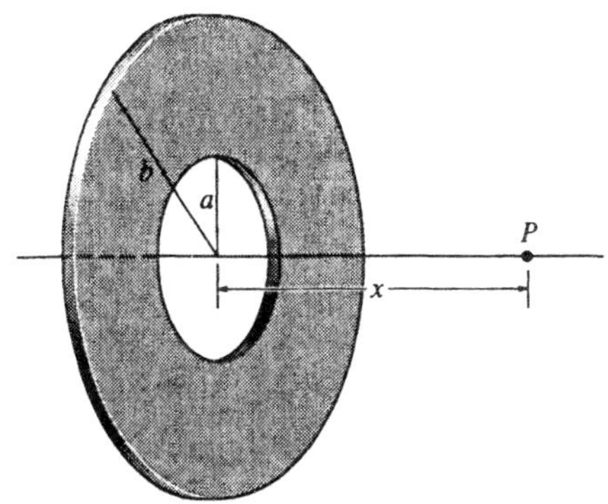

FIGURE 25-40 Problem 39.

Solution

The annulus can be considered to be composed of thin rings of radius $r(a \leq r \leq b)$ and charge $dq = 2\pi\sigma r\ dr$ (see Example 25-7 and Figs. 25-15 and 16). The element of potential from a ring on its axis, a distance x from the center, is $dV = k\ dq/\sqrt{x^2+r^2}$ (see Example 25-6) so the potential from the whole annulus is:

$$V = \int dV = 2\pi\sigma k \int_a^b \frac{r\ dr}{\sqrt{x^2+r^2}} = 2\pi k\sigma \left|\sqrt{x^2+r^2}\right|_a^b$$
$$= 2\pi k\sigma(\sqrt{x^2+b^2} - \sqrt{x^2+a^2}).$$

(Note: This reduces to the potential on the axis of a uniformly charged disk if $a \to 0$.)

Problem

41. (a) Find the potential as a function of position in the electric field $\mathbf{E} = E_0(\hat{\mathbf{\imath}} + \hat{\mathbf{\jmath}})$, where $E_0 =$ 150 V/m. Take the zero of potential at the origin. (b) Find the potential difference from the point $x = 2.0$ m, $y = 1.0$ m to the point $x = 3.5$ m, $y = -1.5$ m.

Solution

(a) Equation 25-2b gives the potential for a uniform field. Take the zero of potential at the origin (point A in Equation 25-2b) and let $\boldsymbol{\ell} = \mathbf{r} = x\hat{\mathbf{\imath}} + y\hat{\mathbf{\jmath}} + z\hat{\mathbf{k}}$ be the vector from the origin to the field point (point B in Equation 25-2b). Then $\Delta V_{AB} = V_B - V_A = V(\mathbf{r}) - 0 = V(x, y) = -E_0(\hat{\mathbf{\imath}} + \hat{\mathbf{\jmath}}) \cdot \mathbf{r} = -E_0(x+y)$. (The potential is independent of z, so we wrote $V(\mathbf{r}) = V(x,\ y)$.)
(b) $V(3.5\text{ m}, -1.5\text{ m}) - V(2.0\text{ m}, 1.0\text{ m}) =$ $-(150\text{ V/m})(3.5\text{ m} - 1.5\text{ m} - 2.0\text{ m} - 1.0\text{ m}) = 150\text{ V}$.

Section 25-4: Potential Difference and the Electric Field

Problem

43. Figure 25-41 shows a plot of potential versus position along the x-axis. Make a plot of the x component of the electric field for this situation.

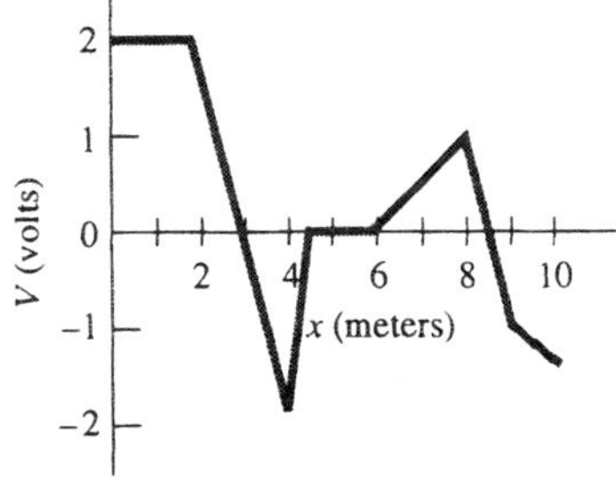

FIGURE 25-41 Problem 43.

Solution

Figure 25-24 illustrates the relation $E_x = -dV/dx$, which may be used to estimate E_x for the seven straight-line segments shown in Fig. 25-41. Thus, for $x = 0$ to 2 m, $E_x = 0$, for $x = 2$ m to 4 m, $E_x =$ $-(-2\text{ V} - 2\text{ V})/(4\text{ m} - 2\text{ m}) = 2\text{ V/m}$; etc., as sketched below.

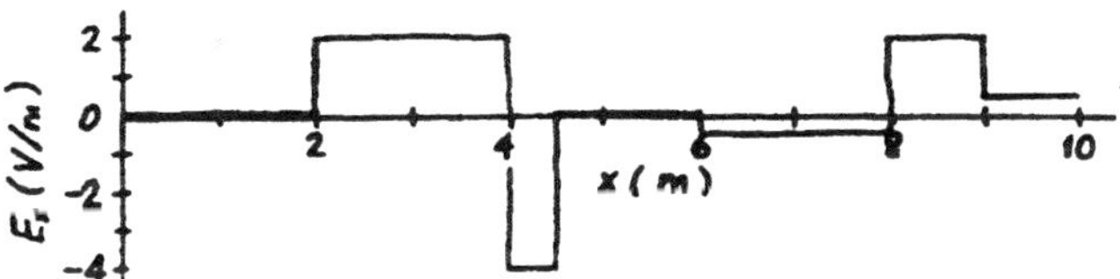

Problem 43 Solution.

Problem

45. The potential in a certain region is given by $V = axy$, where a is a constant. (a) Determine the electric field in the region. (b) Sketch some equipotentials and field lines.

Solution

(a) The x and y components of the electric field can be found from Equation 25-10: $E_x = -\partial V/\partial x =$ $-\partial/\partial x(axy) = -ay$, and $E_y = -\partial V/\partial y = -ax$. Thus $\mathbf{E} = -a(y\hat{\mathbf{\imath}} + x\hat{\mathbf{\jmath}})$. (The field has no z component.)
(b) See sketch below. The field lines (dashed) are perpendicular to the equipotentials (solid) in the direction of decreasing potential (arrow for $a > 0$, in this case). These equipotentials and field lines are confocal hyperbolas, proportional to xy and $\frac{1}{2}(x^2 - y^2)$ respectively, and are sketched only for x and y in the first quadrant.

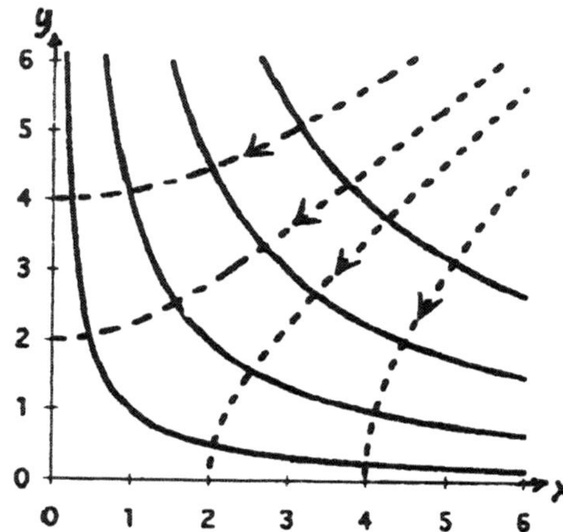

Problem 45 Solution.

Problem

47. The electric potential in a region of space is given by $V = 2xy - 3zx + 5y^2$, with V in volts and the coordinates in meters. If point P is at $x = 1$ m, $y = 1$ m, $z = 1$ m, find (a) the potential at P and (b) the x, y, and z components of the electric field at P.

Solution

(a) Direct substitution gives $V(P) = 2(1)(1) - 3(1)(1) + 5(1)^2 = 4$ V. (b) Use of Equation 25-10 gives $E_x = -\partial V/\partial x = -2y + 3z$, $E_y = -\partial V/\partial y = -2x - 10y$, and $E_z = -\partial V/\partial z = 3x$. At $P(x = y = z = 1)$, $E_x = 1$ V/m, $E_y = -12$ V/m, and $E_z = 3$ V/m.

Problem

49. Use the result of Example 25-6 to determine the on-axis field of a charged ring, and verify that your answer agrees with the result of Example 23-8.

Solution

On the axis of a uniformly charged ring (the x-axis), $V = kQ/\sqrt{x^2 + a^2}$ (Equation 25-9), and the electric field only has an x component (by symmetry). Then $E = (-dV/dx)\hat{\imath} = kQx(x^2 + a^2)^{-3/2}\hat{\imath}$, in accord with Example 23-8. (In general, one needs to know the potential in a 3-dimensional region in order to calculate the field from its partial derivatives.)

Problem

51. The electric potential in a region is given by $V = -V_0(r/R)$, where V_0 and R are constants, r is the radial distance from the origin, and where the zero of potential is taken at $r = 0$. Find the magnitude and direction of the electric field in this region.

Solution

Since $V = -(V_0/R)r$ depends only on r, the field is spherically symmetric and its direction is radial. From Equation 25-10 (which applies unmodified for the radial coordinate), $E_r = -dV/dr = V_0/R$, and $\mathbf{E} = (V_0/R)\hat{\mathbf{r}}$ ($\hat{\mathbf{r}}$ is a unit vector radially outward).

Section 25-5: Potentials of Charged Conductors

Problem

53. The spark plug in an automobile engine has a center electrode made from wire 2.0 mm in diameter. The electrode is worn to a hemispherical shape, so it behaves approximately like a charged sphere. What is the minimum potential on this electrode that will ensure the plug sparks in air? Neglect the presence of the second electrode.

Solution

If we can treat the field from the central electrode as that from an isolated sphere, then $E = kq/R^2$ and $V = kq/R$, so that $V = RE$. Dielectric breakdown in air would occur for potentials exceeding $V = (1 \text{ mm})\times(3\times10^6 \text{ V/m}) = 3$ kV.

Problem

55. Two metal spheres each 1.0 cm in radius are far apart. One sphere carries 38 nC of charge, the other -10 nC. (a) What is the potential on each? (b) If the spheres are connected by a thin wire, what will be the potential on each once equilibrium is reached? (c) How much charge must move between the spheres in order to achieve equilibrium?

Solution

(a) Since the spheres are far apart (approximately isolated), we can use Equation 25-11 to find their potentials: $V_1 = kQ_1/R_1 = (9 \text{ GN}\cdot\text{m}^2/\text{C}^2)(38 \text{ nC})\div(1 \text{ cm}) = 34.2$ kV and $V_2 = kQ_2/R_2 = -9$ kV.
(b) When connected by a thin wire, the spheres reach electrostatic equilibrium with the same potential, so $V = kQ_1'/R_1 = kQ_2'/R_2$. Since the radii are equal, so must be the charges, $Q_1' = Q_2'$. The total charge is $38 \text{ nC} - 10 \text{ nC} = 28 \text{ nC} = Q_1' + Q_2' = 2Q_1'$ (if we assume that the wire is so thin that it has a negligible charge), so $Q_1' = Q_2' = 14$ nC. Then $V' = k(14 \text{ nC})\div(1 \text{ cm}) = 12.6$ kV. (c) In this process, the first sphere loses $38 - 14 = 24$ nC to the second.

Problem

57. Two conducting spheres are each 5.0 cm in diameter and each carries 0.12 μC. They are 8.0 m

apart. Determine (a) the potential on each sphere; (b) the field strength at the surface of each sphere; (c) the potential midway between the spheres; (d) the potential difference between the spheres.

Solution

Since the spheres are small and widely separated, at small distances ($\ll$8 m) they behave like isolated spheres, and at large distances ($\gg$5 cm) they behave like two point charges. (a) At the surface of either sphere, Equation 25-11 gives $V_{\text{surf}} = kq/R =$ $(9 \times 10^9\ \text{N·m}^2/\text{C}^2)(1.2\times10^{-7}\ \text{C})/(0.025\ \text{m}) = 43.2\ \text{kV}$. (b) Then $E_{\text{surf}} = \sigma/\varepsilon_0 = kq/R^2 = V_{\text{surf}}/R =$ $(43.2\ \text{kV})/(2.5\ \text{cm}) = 1.73\times10^6\ \text{V/m}$.
(c) Midway between the spheres, the potential from each one is the same, so $V_{\text{mid-pt.}} = 2kq/r = 2\times$ $(9 \times 10^9\ \text{V·m/C})(1.2\times10^{-7}\ \text{C})/(4\ \text{m}) = 540\ \text{V}$.
(d) The spheres are at the same potential, so the difference is zero.

Paired Problems

Problem

59. Three 50-pC charges sit at the vertices of an equilateral triangle 1.5 mm on a side. How much work would it take to bring a proton from very far away to the midpoint of one of the triangle's sides?

Solution

Two of the charges are at distances of $\frac{1}{2}(1.5\ \text{mm}) =$ 0.75 mm, and the third is at $\sqrt{3}\,(0.75\ \text{mm})$ from the midpoint of one side. Therefore, the potential at this point (Equation 25-6) is $V(P) = k(50\ \text{pC})(1 + 1 + 1/\sqrt{3})/(0.75\ \text{mm}) = 1.55\ \text{kV}$. The work it would take to move a proton of charge e from infinity (where the potential is zero) to this point is $eV = 1.55\ \text{keV} =$ $(1.6\times10^{-19}\ \text{C})(1.55\ \text{kV}) = 2.47\times10^{-16}\ \text{J}$.

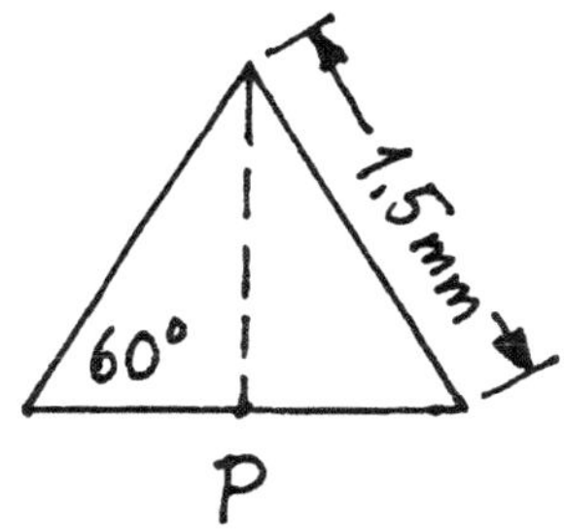

Problem 59 Solution.

Problem

61. A pair of equal charges q lies on the x-axis at $x = \pm a$. (a) Find expressions for the potential at points on the x-axis for which $x > a$ and (b) show that your result reduces to a point-charge potential for $x \gg a$.

Solution

(a) From Equation 25-6, the potential on the x-axis is $V(x) = kq(|x - a|^{-1} + |x + a|^{-1})$. For $x > a$, this becomes $V(x) = kq[(x - a)^{-1} + (x + a)^{-1}] =$ $2kqx(x^2 - a^2)^{-1}$. (b) For $x \gg a, V(x) \to 2kq/x$, the potential of a point charge $2q$ at a distance x.

Problem

63. A 2.0-cm-radius metal sphere carries 75 nC and is surrounded by a concentric spherical conducting shell of radius 10 cm carrying −75 nC. (a) Find the potential difference between the shell and the sphere. (b) How would your answer change if the shell charge were changed to +150 nC?

Solution

(a) The electric field outside the sphere (radius $R_1 =$ 2 cm), but inside the shell inner radius $R_2 = 10$ cm), is only due to the charge q_1 on the sphere (recall Gauss's law), and equals kq_1/r^2 radially outward. The potential difference is $V_1 - V_2 = -\int_{R_2}^{R_1} kq_1\,dr/r^2 =$ $kq_1(R_1^{-1} - R_2^{-1}) = (9\ \text{GN·m}^2/\text{C}^2)(75\ \text{nC})[(2\ \text{cm})^{-1} -$ $(10\ \text{cm})^{-1}] = 27.0\ \text{kV}$. (We used Equation 25-2a, with A at R_2 and B at R_1. Note that $\mathbf{E}\cdot d\mathbf{r} = kq_1 dr/r^2$ regardless of the choice of points A and B.)
(b) Adding more charge to the conducting shell does not affect the field inside, nor the potential difference between points inside, like $V_1 - V_2$ in part (a). The field outside the shell and the potential relative to infinity would change.

Problem

65. On the x-axis, the electric field of a certain charge distribution is given by $\mathbf{E} = a/x^4\hat{\imath}$, where $a = 55\ \text{V·m}^3$. Find the potential difference from the point $x = 1.3$ m to the point $x = 2.8$ m.

Solution

Direct application of Equation 25-2a yields:

$$V(2.8\ \text{m}) - V(1.3\ \text{m})$$
$$= -\int_{1.3\,\text{m}}^{2.8\,\text{m}} E_x\,dx = -\int_{1.3\,\text{m}}^{2.8\,\text{m}} \frac{a}{x^4}\,dx$$
$$= \left(\frac{55\ \text{V·m}^3}{3}\right)\left[\frac{1}{(2.8\ \text{m})^3} - \frac{1}{(1.3\ \text{m})^3}\right]$$
$$= -7.51\ \text{V}.$$

Problem

67. The potential as a function of position in a certain region is given by $V(x) = 3x - 2x^2 - x^3$, with x in meters and V in volts. Find (a) all points on the x-axis where $V = 0$, (b) an expression for the electric field, and (c) all points on the x-axis where $\mathbf{E} = 0$.

Solution

(a) The expression for the potential can be factored into $V(x) = x(x+3)(1-x)$, so $V(x) = 0$ at $x = 0$, 1 m, and -3 m. (b) V is independent of y and z, hence $\mathbf{E} = E_x\hat{\mathbf{i}}$ has only an x component, and $E_x = -dV/dx = 3x^2 + 4x - 3$. (c) $E_x = 0$ for $x = (-2 \pm \sqrt{4+9})/3 = 0.535$ m and -1.87 m.

Supplementary Problem

Problem

69. A conducting sphere 5.0 cm in radius carries 60 nC. It is surrounded by a concentric spherical conducting shell of radius 15 cm carrying -60 nC. (a) Find the potential at the sphere's surface, taking the zero of potential at infinity. (b) Repeat for the case when the shell also carries $+60$ nC.

Solution

The potential difference between the sphere and the shell depends only on the electric field inside the shell (which is due to q_{sphere} only). For a spherically symmetric configuration, this was found to be $V_{\text{sphere}} - V_{\text{shell}} = kq_{\text{sphere}}(R^{-1}_{\text{sphere}} - R^{-1}_{\text{shell}})$ in Problem 63. (Here, R_{shell} is the inner radius of the shell.) Outside the shell, the potential depends on the total charge, which, at the surface of a spherical distribution, was found to be $V_{\text{shell}} - V_\infty = k(q_{\text{sphere}} + q_{\text{shell}})/R_{\text{shell}}$ in Example 25-3. (Here, R_{shell} is the outer radius of the shell, the conducting material of which is assumed to be an equipotential region in electrostatic equilibrium, and $V_\infty = 0$.) Thus, $V_{\text{sphere}} - V_\infty = V_{\text{sphere}} - V_{\text{shell}} + V_{\text{shell}} - V_\infty = kq_{\text{sphere}}(R^{-1}_{\text{sphere}} - R^{-1}_{\text{shell}}) + k(q_{\text{sphere}} + q_{\text{shell}})R^{-1}_{\text{shell}} = kq_{\text{sphere}}R^{-1}_{\text{sphere}} + kq_{\text{shell}}R^{-1}_{\text{shell}}$. (For a thin shell, the inner and outer radii are approximately equal.) (a) In this problem, $q_{\text{sphere}} = -q_{\text{shell}} = 60$ nC, and $R_{\text{sphere}} = \frac{1}{3}R_{\text{shell}} = 5$ cm, so $V_{\text{sphere}} = \left(9\times\frac{6}{5}\right)\times\left(1-\frac{1}{3}\right)$ kV $= 7.2$ kV. (b) If q_{shell} is changed to equal $+q_{\text{sphere}}$, $V_{\text{sphere}} = (9\times\frac{6}{5})(1+\frac{1}{3})$ kV $= 14.4$ kV.

Problem

71. The potential on the axis of a uniformly charged disk at 5.0 cm from the disk center is 150 V; the potential 10 cm from the disk center is 110 V. Find the disk radius and its total charge.

Solution

Combining the given data with the potential in Exercise 25-7, we find $150\text{ V} = 2kQ/a^2(\sqrt{(5\text{ cm})^2 + a^2} - 5\text{ cm})$ and $110\text{ V} = 2kQ/a^2(\sqrt{(10\text{ cm})^2 + a^2} - 10\text{ cm})$. The charge can be eliminated by division,

$$\left(\frac{150}{110}\right) = \frac{\sqrt{1 + (a/5\text{ cm})^2} - 1}{\sqrt{4 + (a/5\text{ cm})^2} - 2}.$$

Several lines of algebra to remove the square roots finally yields $a = (5\text{ cm})\sqrt{105\times209}/52 = 14.2$ cm. We can now solve for Q from either of the first two equations, $Q = (Va^2/2k)[\sqrt{x^2+a^2} - x]^{-1} = 1.67$ nC.

Problem

73. A power line consists of two parallel wires 3.0 cm in diameter spaced 2.0 m apart. If the potential difference between the wires is 4.0 kV, what is the charge per unit length on each wire? The wires carry equal but opposite charges. *Hint:* The wires are far enough apart that they don't greatly affect each other's fields.

Solution

Since the radius of the wires ($a = 1.5$ cm) is much smaller than their separation ($b = 200$ cm), their potential difference can be found from the superposition of the potentials of two (approximately infinite) line charges $\pm\lambda$ (as calculated in Example 25-4). Thus, for any two points between the wires,

$$\begin{aligned} V_2 - V_1 &= (V_2^+ - V_1^+) + (V_2^- - V_1^-) \\ &= \frac{\lambda}{2\pi\varepsilon_0}\ln\left(\frac{r_1}{r_2}\right) + \frac{(-\lambda)}{2\pi\varepsilon_0}\ln\left(\frac{b-r_1}{b-r_2}\right) \\ &= \frac{\lambda}{2\pi\varepsilon_0}\ln\left(\frac{r_1(b-r_2)}{r_2(b-r_1)}\right). \end{aligned}$$

To obtain the potential difference between the wires, we let $r_2 = a$ (the positive wire is at a higher potential) and $r_1 = b - a$. Then

$$\Delta V = \frac{\lambda}{2\pi\varepsilon_0}\ln\frac{(b-a)(b-a)}{a(b-b+a)} = \frac{\lambda}{\pi\varepsilon_0}\ln\left(\frac{b-a}{a}\right).$$

From the given numerical values, we can solve for λ:

$$\begin{aligned} \lambda &= \frac{\pi\varepsilon_0\,\Delta V}{\ln\left(\dfrac{b-a}{a}\right)} = \frac{4000\text{ V}}{(36\times10^9\text{ V}\cdot\text{m/C})\ln\left(\dfrac{200}{1.5} - 1\right)} \\ &= 22.7\text{ nC/m}. \end{aligned}$$

(Incidentally, we have found the capacitance per unit length of a bifilar transmission line, $C/\ell = \pi\varepsilon_0 \div \ln\left(\frac{b}{a} - 1\right)$, in the limit $a \ll b$; see Section 26-4.)

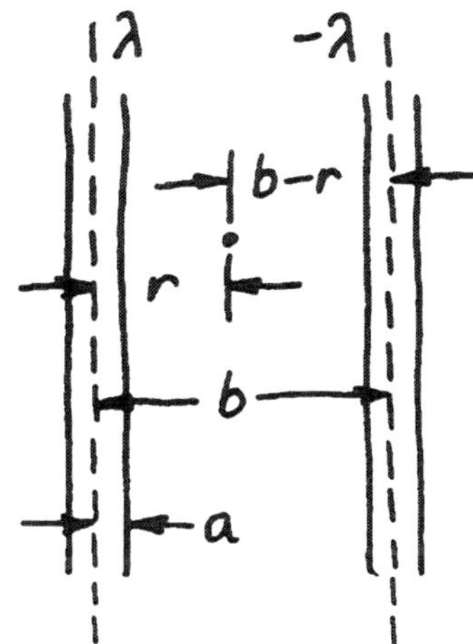

Problem 73 Solution.

Problem

75. A thin rod of length ℓ lies on the x-axis with its center at the origin. It carries a line charge density given by $\lambda = \lambda_0(x/\ell)^2$, where λ_0 is a constant. (a) Find an expression for the potential on the x-axis for $x > \ell/2$. (b) Integrate the charge density to find the total charge on the rod. (c) Show that your answer for (a) reduces to the potential of a point charge whose charge is the answer to (b), for $x \gg \ell$.

Solution

(a) For points P on the x-axis at $x > \ell/2$, the potential from a charge element $dq = \lambda\,dx'$ at x' along the rod $(-\ell/2 \le x' \le \ell/2)$ is $dV = k\,dq/|x - x'| = k\lambda\,dx' \div (x - x')$. (We used x' for the variable position of dq, and took the potential relative to zero at infinity.) The potential due to the entire rod, for $\lambda = \lambda_0(x'/\ell)^2$, is

$$V(x) = \int_{-\ell/2}^{\ell/2} dV = \frac{k\lambda_0}{\ell^2}\int_{-\ell/2}^{\ell/2} \frac{x'^2\,dx'}{(x - x')}$$

$$= \frac{k\lambda_0}{\ell^2}\left|-x^2\ln(x - x') - xx' - \frac{x'^2}{2}\right|_{-\ell/2}^{\ell/2}$$

$$= \frac{k\lambda_0}{\ell^2}\left[x^2\ln\left(\frac{x + \ell/2}{x - \ell/2}\right) - x\ell\right].$$

(Use partial fractions, $x'^2/(x - x') = -x - x' + x^2 \div (x - x')$, or standard tables to evaluate the integral.) (b) The total charge on the rod is $Q = \int dq = \int_{-\ell/2}^{\ell/2}(\lambda_0/\ell^2)x^2\,dx = (\lambda_0/\ell^2)\left(\frac{2}{3}\right)(\ell/2)^3 = \lambda_0\ell/12$. (c) For $x \gg \ell$, we can expand the logarithms, $\ln(1 \pm \ell/2x) = \pm(\ell/2x) - \frac{1}{2}(\ell/2x)^2 \pm\frac{1}{3}(\ell/2x)^3 - \ldots$.

Therefore:

$$V(x) = (k\lambda_0/\ell^2)[x^2\ln(1 + \ell/2x) - x^2\ln(1 - \ell/2x) - x\ell]$$

$$= \left(\frac{k\lambda_0}{\ell^2}\right)\left[x^2\left(\frac{\ell}{2x} - \frac{1}{2}\left(\frac{\ell}{2x}\right)^2 + \frac{1}{3}\left(\frac{\ell}{2x}\right)^3 - \cdots\right) - x^2\left(-\frac{\ell}{2x} - \frac{1}{2}\left(\frac{\ell}{2x}\right)^2 - \frac{1}{3}\left(\frac{\ell}{2x}\right)^3 - \cdots\right) - x\ell\right]$$

$$= \left(\frac{k\lambda_0}{\ell^2}\right)\left[x^2\left(\frac{\ell}{x} + \frac{2}{3}\left(\frac{\ell}{2x}\right)^3 + \cdots\right) - x\ell\right]$$

$$= \frac{k\lambda_0\ell}{12x} = kQ/x,$$

as expected.

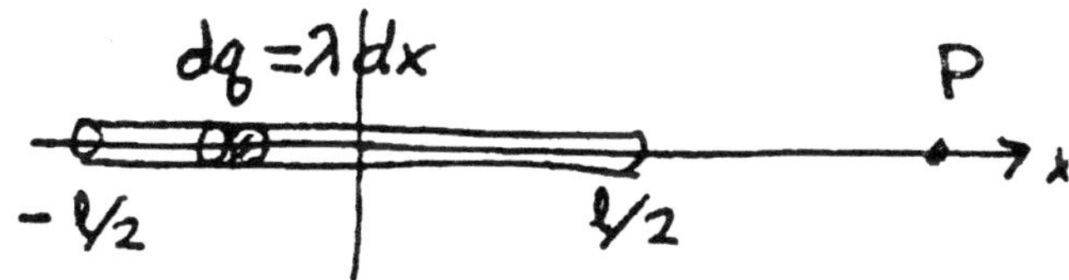

Problem 75 Solution.

Problem

77. For the situation of Example 25-10, find an equation for the equipotential with $V = 0$ in the x-y plane. Plot the equipotential, and show that it passes through the points described in Example 25-10 and its exercise.

Solution

The potential in the x-y plane, for the charges in Example 25-10, is

$$V(x, y) = \frac{k(2q)}{r_+} + \frac{k(-q)}{r_-} = kq\left[\frac{2}{\sqrt{(x + a)^2 + y^2}} - \frac{1}{\sqrt{(x - a)^2 + y^2}}\right],$$

where r_+, r_- are the distances from a point (x, y) to the charges $+2q$ at $(-a, 0)$ and $-q$ at $(a, 0)$ respectively. The potential is zero when $r_+ = 2r_-$, or $(x + a)^2 + y^2 = 4[(x - a)^2 + y^2]$. This equation simplifies to $y^2 + x^2 - 10ax/3 + a^2 = 0$, or after completion of the square in x, to $y^2 + (x - 5a/3)^3 = (4a/3)^2$. This is the equation of a circle centered at $(5a/3, 0)$ with radius $4a/3$. The intercepts of the circle on the x-axis are at $5a/3 \pm 4a/3 = 3a$ or $a/3$, as found in Example 25-10 and the following exercise.

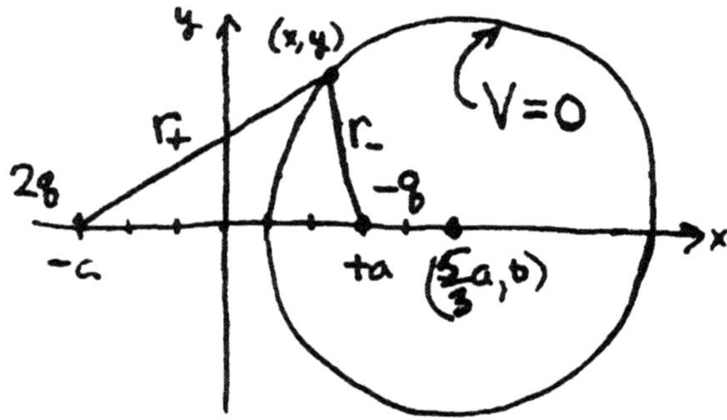

Problem 77 Solution.

Problem

79. An open-ended cylinder of radius a and length $2a$ carries charge q spread uniformly over its surface. Find the potential on the cylinder axis at its center. *Hint:* Treat the cylinder as a stack of charged rings, and integrate.

Solution

The cylinder can be considered to be composed of rings of radius a, width dx, and charge $dq = (q/2a)\, dx$. The potential at the center of the cylinder (which we take as the origin, with x-axis along the cylinder axis) due to a ring at $x\,(-a \leq x \leq a)$ is $dV = k\, dq/\sqrt{x^2 + a^2}$ (see Example 25-6). The whole potential at the center follows from integration:

$$V = \int_{-a}^{a} \left(\frac{kq}{2a}\right) \frac{dx}{\sqrt{x^2 + a^2}} = \frac{kq}{2a} \ln\left(\frac{a + \sqrt{2}a}{-a + \sqrt{2}a}\right)$$

$$= \frac{kq}{a} \ln(1 + \sqrt{2}) = 0.881 \frac{kq}{a}.$$

CHAPTER 26 ELECTROSTATIC ENERGY AND CAPACITORS

Section 26-1: Energy of a Charge Distribution

Problem

1. Three point charges, each of $+q$, are moved from infinity to the vertices of an equilateral triangle of side ℓ. How much work is required?

Solution

The sentence preceding Example 26-1 allows us to rewrite Equation 26-1 (for the electrostatic energy of a distribution of point charges) as $W = \Sigma_{\text{pairs}}\ kq_iq_j/r_{ij}$. For three equal charges (three different pairs) at the corners of an equilateral triangle ($r_{ij} = \ell$ for each pair) $W = 3kq^2/\ell$.

Problem

3. Four 50-μC charges are brought from far apart onto a line where they are spaced at 2.0-cm intervals. How much work does it take to assemble this charge distribution?

Solution

Number the charges $q_i = 50\ \mu\text{C}$, $i = 1, 2, 3, 4$, as they are spaced along the line at $a = 2$ cm intervals. There are six pairs, so $W = \Sigma_{\text{pairs}}\ kq_iq_j/r_{ij} = k(q_1q_2/a + q_1q_3/2a + q_1q_4/3a + q_2q_3/a + q_2q_4/2a + q_3q_4/a) = (kq^2/a)(1 + \frac{1}{2} + \frac{1}{3} + 1 + \frac{1}{2} + 1) = 13kq^2/3a = 13\times(9\times10^9\ \text{m/F})(50\ \mu\text{C})^2/(3\times2\ \text{cm}) = 4.88\ \text{kJ}$. (See solution to Problem 1.)

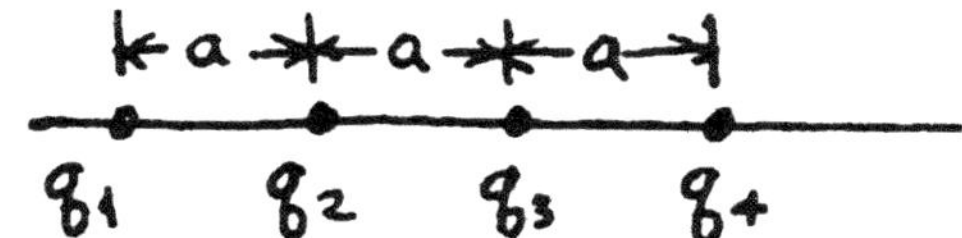

Problem 3 Solution.

Problem

5. Suppose two of the charges in Problem 1 are held in place, while the third is allowed to move freely. If this third charge has mass m, what will be its speed when it's far from the other two charges?

Solution

With one charge removed to infinity, the potential energy is reduced to that of just one pair of charges, $W_f = kq^2/\ell$. The initial potential energy was $W_i = 3kq^2/\ell$ (see Problem 1), so the kinetic energy of the charge at infinity (from the conservation of energy) is $K = W_i - W_f = 2kq^2/\ell$. Thus, $\nu = \sqrt{2K/m} = q/\sqrt{\pi\varepsilon_0 m\ell}$.

Problem

7. Four identical charges q, initially widely separated, are brought to the vertices of a tetrahedron of side a (Fig. 26-26). Find the electrostatic energy of this configuration.

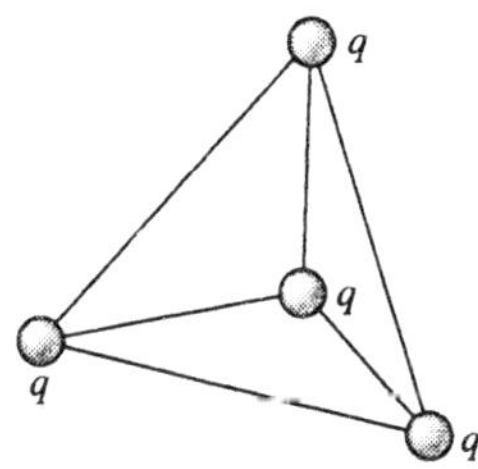

FIGURE 26-26 Problem 7.

Solution

There are six different pairs of equal charges and the separation of any pair is a. Thus, $W = \Sigma_{\text{pairs}}\ kq_iq_j \div a = 6kq^2/a$. (See Problem 1.)

Section 26-2: Two Isolated Conductors

Problem

9. Two square conducting plates 25 cm on a side and 5.0 mm apart carry charges $\pm1.1\ \mu$C. Find (a) the electric field between the plates, (b) the potential difference between the plates, and (c) the stored energy.

Solution

(a) The electric field between two closely spaced, oppositely charged, parallel conducting plates is approximately uniform (directed from the positive to the negative plate), with strength $E = \sigma/\varepsilon_0 = q \div \varepsilon_0 A = (1.1\ \mu\text{C})/(8.85\ \text{pF/m})(0.25\ \text{m})^2 = 1.99\ \text{MV/m}$.

(See the last paragraph of Section 24-6.) (b) Since E is uniform, $V = Ed = (1.99\text{ MV/m})(5\text{ mm}) = 9.94\text{ kV}$. (See Section 26-2.) (c) The energy stored is $U = \frac{1}{2}q^2d/\varepsilon_0 A = \frac{1}{2}qV = \frac{1}{2}(1.1\ \mu\text{C})(9.94\text{ kV}) = 5.47\text{ mJ}$. (See Equation 26-2, and note that $U = \frac{1}{2}\varepsilon_0 E^2 Ad$.)

Problem

11. (a) How much charge must be transferred between the initially uncharged plates of the preceding problem in order to store 15 mJ of energy? (b) What will be the potential difference between the plates?

Solution

(a) From Equation 26-2, $Q = \sqrt{2\varepsilon_0 AU/d} = [2(8.85\text{ pF/m})(5\text{ cm})^2(15\text{ mJ})/(1.2\text{ mm})]^{1/2} = 0.744\ \mu\text{C}$. (b) The argument leading to Equation 26-2 shows that $V = Qd/\varepsilon_0 A = 40.3\text{ kV}$. Alternatively, the second expression for U in the solution to Problem 9 part (c) gives the same results, $V = 2U/Q = 2(15\text{ mJ})/(0.744\ \mu\text{C})$.

Problem

13. A conducting sphere of radius a is surrounded by a concentric spherical shell of radius b. Both are initially uncharged. How much work does it take to transfer charge from one to the other until they carry charges $\pm Q$?

Solution

When a charge q (assumed positive) is on the inner sphere, the potential difference between the spheres is $V = kq(a^{-1} - b^{-1})$. (See the solution to Problem 25-63(a).) To transfer an additional charge dq from the outer sphere requires work $dW = V\,dq$, so the total work required to transfer charge Q (leaving the spheres oppositely charged) is $W = \int_0^Q V\,dq = \int_0^Q kq\,dq(a^{-1} - b^{-1}) = \frac{1}{2}kQ^2(a^{-1} - b^{-1})$. (Incidentally, this shows that the capacitance of this spherical capacitor is $1/k(a^{-1} - b^{-1}) = ab/k(b - a)$; see Equation 26-8a.)

Problem

15. Two conducting spheres of radius a are separated by a distance $\ell \gg a$; since the distance is large, neither sphere affects the other's electric field significantly, and the fields remain spherically symmetric. (a) If the spheres carry equal but opposite charges $\pm q$, show that the potential difference between them is $2kq/a$. (b) Write an expression for the work dW involved in moving an infinitesimal charge dq from the negative to the positive sphere. (c) Integrate your expression to find the work involved in transferring a charge Q from one sphere to the other, assuming both are initially uncharged.

Solution

(a) The potential difference between the two (essentially isolated) spheres is $\Delta V = kq/a - k(-q) \div a = 2kq/a$ (see Equation 25-12). (b) ΔV is the work per unit positive charge transferred between the spheres, so $dW = dq\ \Delta V = 2kq\ dq/a$. (c) The integration yields $W = \int dW = \int_0^Q 2kq\ dq/a = kQ^2/a$.

Problem

17. A car battery stores about 4 MJ of energy. If all this energy were used to create a uniform electric field of 30 kV/m, what volume would it occupy?

Solution

In a uniform field, Equation 26-4 can be written as $U = \frac{1}{2}\varepsilon_0 E^2 \times$ (Volume of field region). Therefore, the volume is $2(4\text{ MJ})/(8.85\text{ pF/m})(30\text{ kV/m})^2 = 1.00\times 10^9\text{ m}^3 = 1\text{ km}^3$.

Problem

19. Find the electric field energy density at the surface of a proton, taken to be a uniformly charged sphere 1 fm in radius.

Solution

For this model of the proton, the field strength at the surface is $E = ke/R^2$ (from spherical symmetry and Gauss's law). Thus, the energy density in the surface electric field is $u = \frac{1}{2}\varepsilon_0 E^2 = ke^2/8\pi R^4 \simeq (9\times 10^9\text{ m/F})(1.6\times 10^{-19}\text{ C})^2/8\pi\ (1\text{ fm})^4 = 9.17\times 10^{30}\text{ J/m}^3 = 57.3\text{ keV/fm}^3$.

Problem

21. The electric field strength as a function of position x in a certain region is given by $E = E_0(x/x_0)$, where $E_0 = 24\text{ kV/m}$ and $x_0 = 6.0\text{ m}$. Find the total energy stored in a cube 1.0 m on a side, located between $x = 0$ and $x = 1.0\text{ m}$. (The field strength is independent of y and z.)

Solution

Since there is no y or z dependence, the volume element of the cube can be written as $dV = \ell^2\ dx$, where $\ell = 1\text{ m}$ is the cube's edge. Then $U = \int u\ dV = \int_0^\ell \frac{1}{2}\varepsilon_0(E_0/x_0)^2 x^2 \ell^2\,dx = \frac{1}{2}\varepsilon_0(E_0/x_0)^2\ell^5/3$. Numerically, $U = \frac{1}{6}(8.85\text{ pF/m})(24\text{ kV/m})^2(1\text{ m})^5 \div (6\text{ m})^2 = 23.6\ \mu\text{J}$.

Problem

23. A sphere of radius R carries a total charge Q distributed over its surface. Show that the total energy stored in its electric field is $U = kQ^2/2R$.

Solution

The calculation of the electrostatic energy for a sphere with uniform surface charge density is, in fact, given in Example 26-3. We simply set $R_2 = R$, the radius of the sphere, and $R_1 = \infty$ (so the integral covers all the space where the field is non-zero).

Problem

25. Two 4.0-mm-diameter water drops each carry 15 nC. They are initially separated by a great distance. Find the change in the electrostatic potential energy if they are brought together to form a single spherical drop. Assume all charge resides on the drops' surfaces.

Solution

The initial electrostatic energy of two isolated spherical drops, with charge Q on their surfaces and radii R, is $U_i = 2(\frac{1}{2}kQ^2/R)$ (see Problem 23 and Example 26-3). Together, a drop of charge $2Q$, radius $2^{1/3}R$, and energy $U_f = \frac{1}{2}k(2Q)^2/(2^{1/3}R) = 2^{2/3}kQ^2/R$, is created. The work required is the difference in energy, $W = U_f - U_i = (2^{2/3} - 1)kQ^2 \div R = (0.587)(9\times10^9 \text{ m/F})(1.5\times10^{-8} \text{ C})^2 \div (2\times10^{-3} \text{ m}) = 5.95\times10^{-4}$ J.

Problem

27. A long, solid rod of radius a carries uniform volume charge density ρ. Find an expression for the electrostatic energy per unit length contained *within* the rod. *Hint:* See Problem 24-31.

Solution

The electric field within such a rod (assumed to have line symmetry) is radially away from the axis with magnitude $E_r = \rho r/2\varepsilon_0$ (see Problem 24-31). The energy density on a cylindrical shell of radius r, length ℓ, and volume $dV = 2\pi r\ell\, dr$, is $u = \frac{1}{2}\varepsilon_0 E_r^2 = \rho^2 r^2/8\varepsilon_0$. Hence, the energy per unit length inside the rod is $U/\ell = \ell^{-1}\int_{\text{rod}} u\, dV = (\pi\rho^2/4\varepsilon_0)\int_0^a r^3\, dr = \pi\rho^2 a^4/16\varepsilon_0$.

Problem

29. The "memory" capacitor in a VCR has a capacitance of 4.0 F and is charged to 3.5 V. What is the charge on its plates?

Solution

The definition of capacitance (Equation 26-5) gives the magnitude of the charge on either plate, $Q = CV = 4.0 \text{ F} \times 3.5 \text{ V} = 14.0$ C. (This is a very large capacitor.)

Problem

31. Figure 26-27 shows data from an experiment in which known amounts of charge are placed on a capacitor and the resulting voltage measured. Fit a line to the data, and use it to determine the capacitance.

Solution

Since $V = Q/C$, the slope of the best straight line through the data points is the inverse of the capacitance, or $C = (Q_2 - Q_1)/(V_2 - V_1)$. An "eyeball fit" to Fig. 26-27 passes through the origin and (12 mC, 1.85 V), so $C \approx 12 \text{ mC}/1.85 \text{ V} \approx 6.5$ mF.

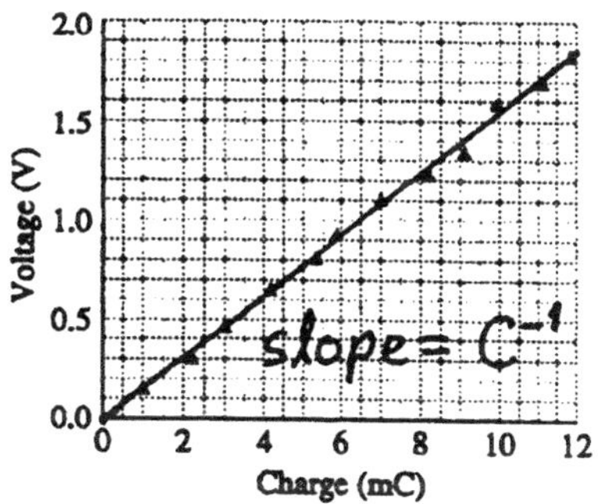

FIGURE 26-27 Problem 31 Solution.

Problem

33. Find the capacitance of a parallel-plate capacitor consisting of circular plates 20 cm in radius separated by 1.5 mm.

Solution

For a (closely spaced) parallel plate capacitor, with circular plates, Example 26-4 shows that $C = \varepsilon_0\pi r^2 \div d = (8.85 \text{ pF/m})\pi(20 \text{ cm})^2/(1.5 \text{ mm}) = 741$ pF.

Problem

35. Find the capacitance of a 1.0-m-long piece of coaxial cable whose inner conductor radius is 0.80 mm and whose outer conductor radius is 2.2 mm, with air in between.

Solution

The capacitance of air-filled ($\kappa = 1$) cylindrical capacitor was found in Example 26-5: $C = 2\pi\varepsilon_0\ell \div \ln(b/a) = 2\pi(8.85 \text{ pF/m})(1 \text{ m})/\ln(2.2/0.8) = 55.0$ pF.

Problem

37. Figure 26-28 shows a capacitor consisting of two electrically connected plates with a third plate between them, spaced so its surfaces are a distance d from the other plates. The plates have area A. Neglecting edge effects, show that the capacitance is $2\varepsilon_0 A/d$.

FIGURE 26-28 Problem 37.

Solution

When the third (middle) plate is positively charged, the electric field (not near an edge) is approximately uniform and away from the plate, with magnitude $E = \sigma/\varepsilon_0$. Since half of the total charge Q is on either side (by symmetry), $\sigma = Q/2A$. The potential difference between the third plate and the outer two plates (which are both at the same potential and carry charges of $-Q/2$ on their inner surfaces) is $V = Ed = \sigma d/\varepsilon_0 = Qd/2\varepsilon_0 A$. Therefore the capacitance is $C = Q/V = 2\varepsilon_0 A/d$. (The arrangement is like two capacitors in parallel.)

Problem

39. Find the capacitance of a capacitor that stores 350 μJ when the potential difference across its plates is 100 V.

Solution

Equation 26-8b relates the capacitance, voltage, and energy stored in a capacitor, so $C = 2U_C/\text{V}^2 = 2(350\ \mu\text{J})/(100\ \text{V})^2 = 70$ nF.

Problem

41. Which can store more energy, a 1-μF capacitor rated at 250 V or a 470 pF capacitor rated at 3 kV?

Solution

The first capacitor stores $U_C = \frac{1}{2}(1\ \mu\text{F})(250\ \text{V})^2 = 31.3$ m of energy, while the second only $\frac{1}{2}(470\ \text{pF})\times(3\ \text{kV})^2 = 2.12$ mJ, about 14.8 times less. (See Equation 26-8b.)

Problem

43. A 0.01-μF, 300-V capacitor costs 25¢, a 0.1-μF, 100-V capacitor costs 35¢, and a 30-μF, 5-V capacitor costs 88¢. (a) Which can store the most charge? (b) Which can store the most energy? (c) Which is the most effective energy storage device, as measured by energy stored per unit cost?

Solution

(a) $Q = CV = (0.01\ \mu\text{F})(300\ \text{V}) = 3\ \mu\text{C}$ for the first capacitor, 10 μC for the second, and 150 μC for the third. (b) $U = \frac{1}{2}QV$ (or $\frac{1}{2}$ CV^2) $= \frac{1}{2}(3\ \mu\text{C})(300\ \text{V}) = 450\ \mu\text{J}$ for the first, 500 μJ for the second, and 375 μJ for the third. (c) The cost effectiveness, measured in $J/¢$, is 18.0, 14.3, and 4.26 for these capacitors, respectively.

Problem

45. A camera flashtube requires 5.0 J of energy per flash. The flash duration is 1.0 ms. (a) What is the power used by the flashtube *while it is actually flashing*? (b) If the flashtube operates at 200 V, what size capacitor is needed to supply the flash energy? (c) If the flashtube is fired once every 10 s, what is its *average* power consumption?

Solution

(a) $\mathcal{P}_{\text{flash}} = W/t = 5\ \text{J}/1\ \text{ms} = 5\ \text{kW}$. (b) $U = \frac{1}{2}CV^2$, so $C = 2U/V^2 = 2(5\ \text{J})/(200\ \text{V})^2 = 250\ \mu\text{F}$. (c) $\mathcal{P}_{\text{av}} = 5\ \text{J}/10\ \text{s} = 0.5$ W, only 10^{-4} times $\mathcal{P}_{\text{flash}}$.

Problem

47. A solid conducting slab is inserted between the plates of a charged capacitor, as shown in Fig. 26-29. The slab thickness is 60% of the plate spacing, and its area is the same as the plates. (a) What happens to the capacitance? (b) What happens to the stored energy, assuming the capacitor is not connected to anything?

Solution

(a) The charge on the plates remains the same, and so does the electric field ($E = \sigma/\varepsilon_0$) in the gaps between either plate and the slab. However, the separation (i.e., the thickness of the field region) between the plates is reduced to 40% of its original value $d' = d_1 + d_2 = 0.4d$, therefore the capacitance is increased, $C' = \varepsilon_0 A/d' = \varepsilon_0 A/0.4d = 2.5$ C. (The equations $V = E\ell$ and $C = Q/V$ lead to the same result.) In fact, the configuration behaves like a series combination of two parallel plate capacitors, $1/C' = C_1^{-1} + C_2^{-1} = (d_1/\varepsilon_0 A) + (d_2/\varepsilon_0 A) = (d_1 + d_2)/\varepsilon_0 A = 0.4d/\varepsilon_0 A =$

1/2.5 C. (b) When the charge is constant (no connections to anything isolates the system), the energy stored is inversely proportional to the capacitance, $U = Q^2/2C$. Thus $U' = Q^2/2C' = Q^2 \div 2(2.5C) = 0.4U$, or the energy decreases to 40% of its original value. (With the slab inserted, there is less field region and less energy stored. While the slab is being inserted, work is done by electrical forces to conserve energy.)

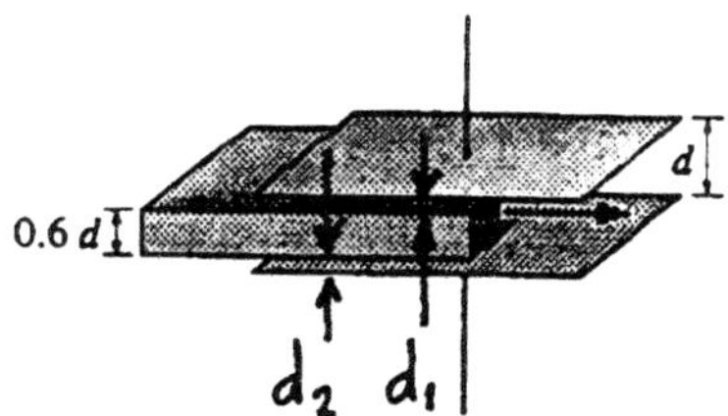

FIGURE 26-29 Problem 47 Solution.

Problem

49. The cylindrical capacitor of Example 26-5 is charged to a voltage V. Obtain an expression for the energy density as a function of radial position in the capacitor, and integrate to show explicitly that the stored energy is $\frac{1}{2}CV^2$.

Solution

The electric field in the capacitor is approximately $\mathbf{E} = \lambda\hat{\mathbf{r}}/2\pi\varepsilon_0 r$, where $\hat{\mathbf{r}}$ is the radial unit vector in cylindrical coordinates (see Example 25-4). (The assumption of line symmetry neglects fringing fields at the ends of the capacitor.) The energy density is $u = \frac{1}{2}\varepsilon_0 E^2 = \lambda^2/8\pi^2\varepsilon_0 r^2$. We can take the volume element to be a cylindrical shell of radius r, thickness dr, and length L, so $dV = 2\pi rL\ dr$. Then the stored energy is

$$U = \int u\ dV = \int_a^b \frac{\lambda^2 2\pi rL\ dr}{8\pi^2\varepsilon_0 r^2} = \frac{\lambda^2 L}{4\pi\varepsilon_0}\int_a^b \frac{dr}{r}$$
$$= \frac{\lambda^2 L}{4\pi\varepsilon_0}\ln\left(\frac{b}{a}\right).$$

Reference to Example 26-5 shows that this is precisely $\frac{1}{2}CV^2$.

Problem

51. Two capacitors are connected in series and the combination charged to 100 V. If the voltage across each capacitor is 50 V, how do their capacitances compare?

Solution

For capacitors in series, the total voltage is the sum of the voltages across each one, $V = V_1 + V_2$, whereas the charge on each capacitor is the same, $Q_1 = Q_2 = C_1V_1 = C_2V_2$. Thus, $V = V_1 + (C_1/C_2)V_1$, or $V_1 = VC_2/(C_1 + C_2)$, and similarly $V_2 = VC_1/(C_1 + C_2)$ (a general result). If $V_1 = V_2 = \frac{1}{2}V$ as in this problem, either equation implies $C_1 = C_2$.

Problem

53. You're given three capacitors: 1.0 μF, 2.0 μF, and 3.0 μF. Find (a) the maximum, (b) the minimum, and (c) two intermediate values of capacitance you could achieve with various combinations of all three capacitors.

Solution

The capacitors can be connected (a) all in parallel: $1 + 2 + 3 = 6\ \mu\text{F}$; (b) all in series: $1/1 + 1/2 + 1/3 = 11/6$, or $6/11 = 0.545\ \mu\text{F}$; (c) one in parallel with the other two in series:

$$1 + \frac{2\times3}{2+3} = \frac{11}{5} = 2.20\ \mu\text{F},$$
$$2 + \frac{1\times3}{1+3} = \frac{11}{4} = 2.75\ \mu\text{F},$$
$$3 + \frac{1\times2}{1+2} = \frac{11}{3} = 3.67\ \mu\text{F},$$

or one in series with the other two in parallel:

$$\frac{1(2+3)}{1+2+3} = \frac{5}{6} = 0.933\ \mu\text{F},$$
$$\frac{2(1+3)}{2+1+3} = \frac{4}{3} = 1.33\ \mu\text{F},$$
$$\frac{3(1+2)}{3+1+2} = \frac{3}{2} = 1.50\ \mu\text{F}.$$

Problem

55. You have an unlimited supply of 2.0-μF, 50-V capacitors. Describe combinations that would be equivalent to (a) a 2.0-μF, 100-V capacitor and (b) a 0.50-μF, 200-V capacitor.

Solution

In parallel, the voltage across each element is the same, so to increase the voltage rating of a combination of equal capacitors, series connections

Problem 55 Solution.

must be considered. The general result of Problem 51 shows that for two equal capacitors in series, the voltage across each is one half the total, so the voltage rating of a series combination is doubled. Thus, in part (a) we must use two capacitors in series (rating 50 V + 50 V = 100 V, and capacitance $C = (2\ \mu\text{F})\times(2\ \mu\text{F})/(2\ \mu\text{F} + 2\ \mu\text{F}) = 1\ \mu\text{F}$), while in part (b), four capacitors in series are required (rating 200 V, and capacitance $C^{-1} = 4(2\ \mu\text{F})^{-1}$ or $C = \frac{1}{2}\ \mu\text{F}$). In part (b), one series combination of four capacitors is sufficient, but in part (a), we need to increase the total capacitance to twice that of just two in series, without altering the voltage rating. This can be accomplished with a parallel combination of two pairs in series, i.e., a parallel combination of two 1 μF, 100 V series pairs. (Note that for equal capacitor elements, a parallel combination of two pairs in series has the same properties as a series combination of two pairs in parallel.) Schematically the connections described look like the following.

Problem

57. What is the equivalent capacitance in Fig. 26-32?

Solution

Number the capacitors as shown. Relative to points A and B, C_1, C_4 and the combination of C_2 and C_3 are in series, so the capacitance is given by $C_{AB}^{-1} = C_1^{-1} + C_4^{-1} + C_{23}^{-1}$. C_{23} is a parallel combination, hence $C_{23} = C_2 + C_3$, therefore $C_{AB}^{-1} = (3\ \mu\text{F})^{-1} + (2\ \mu\text{F})^{-1} + (2\ \mu\text{F} + 1\ \mu\text{F})^{-1}$, or $C_{AB} = \frac{6}{7}\ \mu\text{F} = 0.857\ \mu\text{F}$.

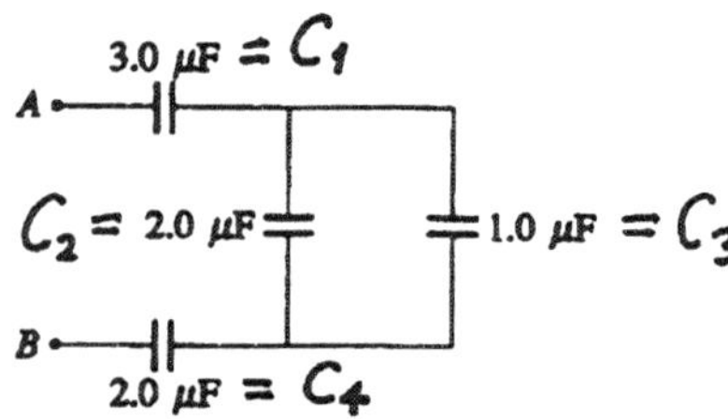

FIGURE 26-32 Problem 57 Solution.

Problem

59. Two capacitors C_1 and C_2 are in series, with a voltage V across the combination. Show that the voltages across the individual capacitors are

$$V_1 = \frac{C_2 V}{C_1 + C_2} \quad \text{and} \quad V_2 = \frac{C_1 V}{C_1 + C_2}.$$

Solution

This is shown in the solution to Problem 51.

Problem

61. A variable "trimmer" capacitor used to make fine adjustments has a capacitance range from 10 to 30 pF. The trimmer is in parallel with a capacitor of about 0.001 μF. Over what percentage range can the capacitance of the combination be varied?

Solution

"Capacitors in parallel add" (Equation 26-9a), so the combination covers a range from 1010 to 1030 pF, or about $\pm 10/1020 \approx \pm 1\%$ from the central value.

Problem

63. A 5.0-μF capacitor is charged to 50 V, and a 2.0-μF capacitor is charged to 100 V. The two are disconnected from their charging batteries and connected in parallel, positive to positive.
(a) What is the common voltage across each after they are connected? *Hint:* Charge is conserved.
(b) Compare the total electrostatic energy before and after the capacitors are connected. Speculate on the discrepancy.

Solution

(a) The charge on the parallel combination is the sum of the original charges, $Q_\parallel = Q_1 + Q_2 = C_1V_1 + C_2V_2 = (5\ \mu\text{F})(50\ \text{V}) + (2\ \mu\text{F})(100\ \text{V}) = 450\ \mu\text{C}$, while the capacitance is $C_\parallel = C_1 + C_2 = 7\ \mu\text{F}$. Thus, the voltage is $V_\parallel = Q_\parallel/C_\parallel = 450\ \mu\text{C}/7\ \mu\text{F} = 64.3\ \text{V}$.
(b) The total energy stored in both capacitors before they are connected is $\frac{1}{2}C_1V_1^2, +\frac{1}{2}C_2V_1^2 = \frac{1}{2}(5\ \mu\text{F})(50\ \text{V})^2 + \frac{1}{2}(2\ \mu\text{F})(100\ \text{V})^2 = 16.3\ \text{mJ}$. After the connection, $U_\parallel = \frac{1}{2}C_\parallel V_\parallel^2 = \frac{1}{2}(7\ \mu\text{F})(64.3\ \text{V})^2 = 14.5\ \text{mJ}$, a difference of 1.79 mJ. It takes work to redistribute the original charges when the capacitors are connected. (The new charges are $Q_1' = (5\ \mu\text{F})(64.3\ \text{V}) = 321\ \mu\text{C}$, and $Q_2' = 129\ \mu\text{C}$, respectively.)

Problem

65. A 470-pF capacitor consists of two circular plates 15 cm in radius, separated by a sheet of polystyrene. (a) What is the thickness of the sheet? (b) What is the working voltage?

Solution

(a) With reference to Equations 26-6, 26-11, and Table 26-1, one finds that $C = \kappa C_0 = \kappa\varepsilon_0 A/d$, or $d = \kappa\varepsilon_0 A/C = (2.6)(8.85\ \text{pF/m})\pi(0.15\ \text{m})^2/470\ \text{pF} = 3.46\ \text{mm}$. (Since this is much less than the radius of the plates, the parallel plate approximation (plane symmetry) is a good one.) (b) The dielectric breakdown field for polystyrene is $E_{\text{max}} = 25\ \text{kV/mm}$, so the maximum voltage for this capacitor is $V_{max} = E_{max}d = (25\text{ kV/mm})(3.46\text{ mm}) = 86.5\text{ kV}$. (Note: in practice, the working voltage would be less than this by a comfortable safety margin.)

Problem

67. Repeat Problem 35 for the more realistic case of a cable insulated with polyethylene.

Solution

With polyethylene insulation occupying the space between the conductors, instead of air, the capacitance is increased by a factor of $\kappa = 2.3$ (see Equation 26-11 and Table 26-1). Thus, $C = \kappa C_0 = (2.3)(55.0\text{ pF}) = 126\text{ pF}$, where we used the value of C_0 from the solution to Problem 35.

Problem

69. The capacitor of the preceding problem is connected to its 900-V charging battery and left connected as the plexiglass sheet is inserted, so the potential difference remains at 900 V. What are (a) the charge on the plates and (b) the stored energy both before and after the plexiglass is inserted?

Solution

(a) The capacitances before and after the insertion of the plexiglass insulation are $C_0 = \varepsilon_0 A/d = (8.85\text{ pF/m})(76\text{ cm}^2)/(1.2\text{ mm}) = 56.1\text{ pF}$, and $C = \kappa C_0 = (3.4)(56.1\text{ pF}) = 191\text{ pF}$, as found previously. Therefore, since the voltage stays at 900 V in this case (due to the battery), $Q_0 = C_0(900\text{ V}) = 50.4\text{ nC}$, and $Q = C(900\text{ V}) = \kappa Q_0 = 172\text{ nC}$, before and after insertion, respectively. (b) The stored energy is $U_0 = \frac{1}{2}C_0(900\text{ V})^2 = 22.7\ \mu\text{J}$ before, and $U = \frac{1}{2}C(900\text{ V})^2 = \kappa U_0 = 77.2\ \mu\text{J}$ after. (The difference between this situation and the one in the previous problem is that the battery does additional work moving more charge to the capacitor plates, while maintaining the constant voltage. Equation 26-12 applies to an isolated capacitor only.)

Paired Problems

Problem

71. A pair of parallel conducting plates of area 0.025 m^2 carrying equal but opposite charges stores 1.6 J in its electric field. When the magnitude of the charge on both plates is increased by 5.0 μC, the stored energy increases to 2.4 J. Find the plate separation.

Solution

Using Equation 26-8a, we can write $U_1 = Q_1^2/2C = 1.6\text{ J}$, and $U_2 = (Q_1 + 5\ \mu\text{C})^2/2C = 2.4\text{ J}$, from which Q_1 can be eliminated and C can be found: $\sqrt{2CU_2} - \sqrt{2CU_1} = 5\ \mu\text{C}$, or $C = [5\ \mu\text{C}/(\sqrt{2\times2.4\text{ J}} - \sqrt{2\times1.6\text{ J}})]^2 = 155\text{ pF}$. For an air-insulated parallel plate capacitor, Equation 26-6 then gives $d = \varepsilon_0 A/C = (8.55\text{ pF/m})(0.025\text{ m}^2)/155\text{ pF} = 1.43\text{ mm}$.

Problem

73. A 20-μF air-insulated parallel-plate capacitor is charged to 300 V. The capacitor is then disconnected from the charging battery, and its plate separation is doubled. Find the stored energy (a) before and (b) after the plate separation increases. Where does the extra energy come from?

Solution

(a) Initially, the stored energy is $U_0 = \frac{1}{2}C_0V_0^2 = \frac{1}{2}(20\ \mu\text{F})(300\text{ V})^2 = 0.9\text{ J}$. (b) Disconnected from the battery, the charge stays constant, but the capacitance is halved when the separation is doubled ($C = \varepsilon_0 A \div 2d = C_0/2$). Therefore, the stored energy is doubled, since $U = Q^2/2C = Q^2/2(C_0/2) = 2U_0 = 1.8\text{ J}$. Work must be done, against the attractive force between the oppositely charged plates, to increase their separation.

Problem

75. In the capacitor network of Fig. 26-33, take $C = 6.0\ \mu\text{F}$. Find (a) the equivalent capacitance between A and B and (b) the charge on C when 30 V is applied between A and B.

Solution

(a) The 2 μF capacitor is in series with the parallel combination of the 1 μF capacitor and the series combination of the 3 μF and 6 μF capacitors (see the numbering added to the figure). Therefore, the total capacitance between A and B is

$$C_{tot} = \frac{C_1C_{\|}}{C_1 + C_{\|}} = \frac{C_1(C_2 + C_{34})}{C_1 + C_2 + C_{34}}$$

$$= \frac{C_1\left(C_2 + \dfrac{C_3C_4}{C_3 + C_4}\right)}{C_1 + C_2 + \dfrac{C_3C_4}{C_3 + C_4}}$$

$$= \frac{2\ \mu\text{F}\left(1\ \mu\text{F} + \dfrac{3\ \mu\text{F}\times 6\ \mu\text{F}}{3\ \mu\text{F} + 6\ \mu\text{F}}\right)}{2\ \mu\text{F} + \left(1\ \mu\text{F} + \dfrac{3\times 6\ \mu\text{F}}{3+6}\right)} = \frac{2\times 3}{5}\ \mu\text{F}$$

$$= 1.2\ \mu\text{F}.$$

(b) When 30 V is applied across A and B, the voltage across the parallel combination (whose capacitance is $1\ \mu\text{F} + 3\times6\ \mu\text{F}/(3+6) = 3\ \mu\text{F}$) is $(30\ \text{V})2/(2+3) = 12\ \text{V}$ (since this is in series with the $2\ \mu\text{F}$ capacitor—see the result of Problem 59). A second application of this result (i.e., Problem 59) to the series combination of the $3\ \mu\text{F}$ and $6\ \mu\text{F}$ capacitor gives $V = (12\ \text{V})3/(3+6) = 4\ \text{V}$ for the voltage across C. Then $Q = CV = (6\ \mu\text{F})(4\ \text{V}) = 24\ \mu\text{C}$.

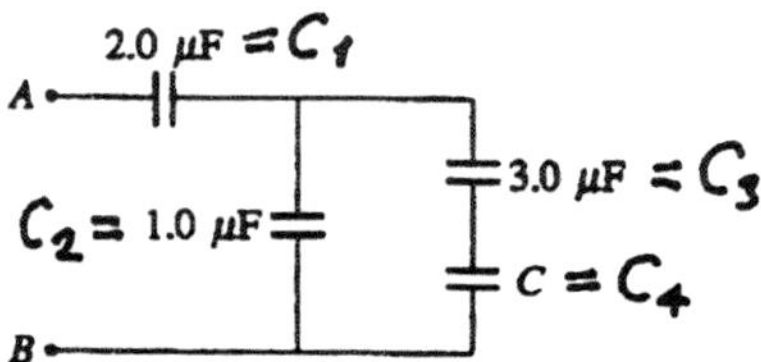

FIGURE 26-33 Problem 75 Solution.

Supplementary Problems

Problem

77. A typical lightning flash transfers 30 C across a potential difference of 30 MV. Assuming such flashes occur every 5 s in the thunderstorm of Example 26-2, roughly how long could the storm continue if its electrical energy were not replenished?

Solution

The energy in the thunderstorm of Example 26-2 was about 1.4×10^{11} J, while the energy in a lightning flash is $qV = (30\ \text{C})(30\ \text{MV}) = 9\times10^{8}$ J. Thus, there is energy for about $1.4\times10^{11}/9\times10^{8} = 156$ flashes, which at a rate of one flash in 5 s, would last for $156\times5\ \text{s} = 13\,\text{min}$.

Problem

79. Six charges $\pm q$, initially widely separated, are positioned to form a hexagon of side a, as shown in Fig. 26-35. What is the electrostatic energy of this configuration?

Solution

The electrostatic energy is $U = \Sigma_{\text{pairs}}\ kq_iq_j/r_{ij}$ (see solution to Problem 1). For the six charges at the corners of a regular hexagon, of side a, shown in Fig. 26-32, there are a total of 15 different pairs: 6 pairs of opposite charges separated by distance a, 6 pairs of equal charges separated by $\sqrt{3}a$, and 3 pairs of opposite charges separated by $2a$ (see geometry added to Fig. 26-32). Thus,

$$U = kq^2\left(-\frac{6}{a} + \frac{6}{\sqrt{3}a} - \frac{3}{2a}\right) = \frac{kq^2}{a}\left(2\sqrt{3} - \frac{15}{2}\right)$$
$$= -4.04\frac{kq^2}{a}.$$

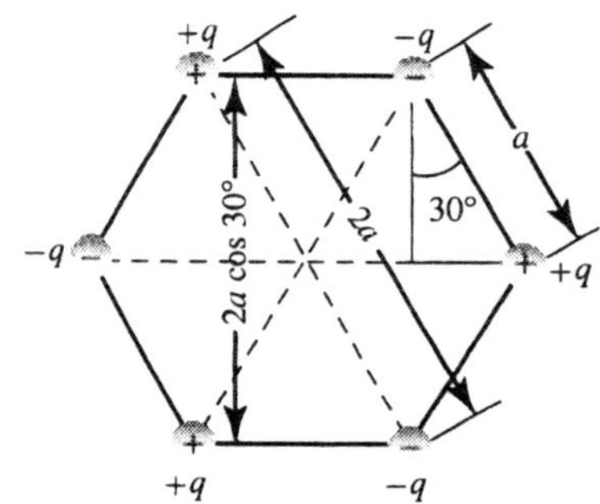

FIGURE 26-35 Problem 79 Solution.

Problem

81. An air-insulated parallel-plate capacitor of capacitance C_0 is charged to voltage V_0 and then disconnected from the charging battery. A slab of material with dielectric constant κ, whose thickness is essentially equal to the capacitor spacing, is then inserted halfway into the capacitor (Fig. 26-36). Determine (a) the new capacitance, (b) the stored energy, and (c) the force on the slab in terms of C_0, V_0, κ, and the capacitor plate length L.

Solution

(a) In so far as fringing fields can be neglected, the electric field between the plates is uniform, $E = V/d$ (but when the dielectric is inserted, $V \neq V_0$ and E depends on x). In fact, on the left side, where the slab has penetrated, $E = (1/\kappa)(\sigma_\ell/\varepsilon_0)$, and on the right, $E = \sigma_r/\varepsilon_0$, where σ_ℓ and σ_r are the charge densities on the left and right sides. Thus, $\sigma_\ell = \kappa\varepsilon_0 E$ and $\sigma_r = \varepsilon_0 E$, and the charge can be written (in terms of geometrical variables superposed on Fig. 26-36) as $q = \sigma_\ell wx + \sigma_r w(L-x) = \varepsilon_0 Ew(\kappa x + L - x) = \varepsilon_0(V/d)w\times(\kappa x + L - x)$. From Equation 26-5, $C = q/V = C_0(\kappa x + L - x)/L$, where $C_0 = \varepsilon_0 A/d$ and $A = Lw$. Although the question specifies $x = \frac{1}{2}L$, for which value the capacitance is $\frac{1}{2}C_0(\kappa+1)$, we give C as a function of x, because we will need to differentiate with respect to x in part (c). (b) When the battery is disconnected, the capacitor is isolated and the charge on it is a constant, $q = q_0$. The stored energy is (Equation 26-8a) $U = q^2/2C = q_0^2L/2C_0(\kappa x + L - x) = U_0L/(\kappa x + L - x)$, where $U_0 = \frac{1}{2}q_0^2/C_0 = \frac{1}{2}C_0V_0^2$. For

$x = \frac{1}{2}L$, the energy is $C_0V_0^2/(\kappa+1)$. (c) The force on a part of an isolated system is related to the potential energy of the system by Equation 8-9. The force on the slab is therefore

$$F_x = -\frac{dU}{dx} = -\frac{d}{dx}\left(\frac{U_0L}{\kappa x + L - x}\right) = \frac{U_0L(\kappa-1)}{(\kappa x + L - x)^2},$$

in the direction of increasing x (so as to pull the slab into the capacitor). For $x = \frac{1}{2}L$, the magnitude of the force is $2C_0V_0^2(\kappa-1)/L(\kappa+1)^2$. It turns out that if we rewrite the force, for any value of x, in terms of the voltage for that x, using $q_0 = C_0V_0 = CV = C_0V(\kappa x + L - x)/L$, the expression can be used in the succeeding problem. Thus,

$$F_x = \frac{C_0V_0^2L(\kappa-1)}{2(\kappa x + L - x)^2} = \frac{C_0}{2}\left(\frac{V}{L}\right)^2 L(\kappa-1)$$
$$= \frac{C_0V^2(\kappa-1)}{2L}.$$

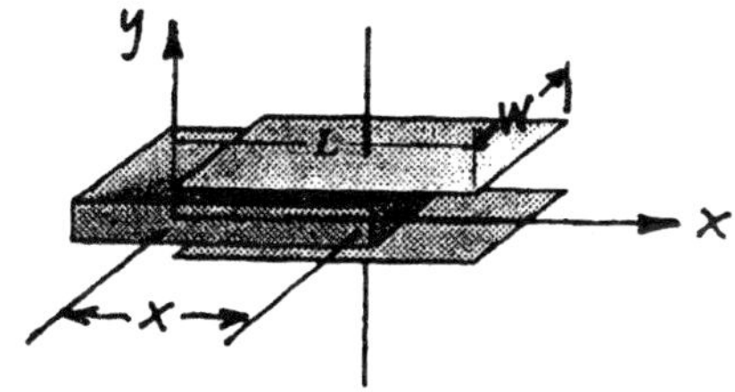

FIGURE 26-36 Problems 81 and 82 Solution.

Problem

83. We live inside a giant capacitor! Its plates are Earth's surface and the ionosphere, a conducting layer of the atmosphere beginning at about 60 km altitude. (a) What is its capacitance? *Hint:* You can treat it as either a spherical or a parallel-plate capacitor. Why? (b) The potential difference between Earth and ionosphere is about 6 MV. Find the total energy stored in this planetary capacitor.

Solution

(a) Since the radius of the Earth, $R_E = 6370$ km $= a$, is much larger than the altitude of the ionosphere, 60 km $= b - a = d$ (which is the separation of the plates in this planetary capacitor), the result of Problem 80 shows that either the spherical or parallel plate expressions for the capacitance are approximately the same. Thus, $C \approx \varepsilon_0 4\pi(6370 \text{ km})^2 \div 60 \text{ km} = 75.2$ mF. (b) Then $U = \frac{1}{2}CV^2 = \frac{1}{2}(75.2 \text{ mF})(6 \text{ MV})^2 = 1.35\times10^{12}$ J.

Problem

85. Equation 26-2 gives the potential energy of a pair of oppositely charged plates. (a) Differentiate this expression with respect to the plate spacing to find the magnitude of the attractive force between the plates. (b) Compare with the answer you would get by multiplying one plate's charge by the electric field between the plates. Why do your answers differ? Which is right?

Solution

(a) Equation 26-2 gives the potential energy of two isolated oppositely charged plates, $U(x) = Q^2x/2\varepsilon_0A$, where x is their separation. Equation 8-9, $F_x = -dU/dx$, implies an attractive force of $F_x = -Q^2 \div 2\varepsilon_0A$ acting between the plates. (b) Multiplying the charge by the total electric field between the plates gives one $Q(\sigma/\varepsilon_0) = Q^2/\varepsilon_0A$, an expression equal to twice the magnitude of the force. The total field includes the field of both plates, whereas the force on one plate depends on only the field of the other plate. (In general, the force per unit area on the surface charge distribution on a conductor is $\frac{1}{2}\sigma E = \sigma^2/2\varepsilon_0$ for the same reason.)

Problem

87. A small dipole lies on the x axis, centered at the origin. Find an expression for the total electrostatic energy contained in a thin cylindrical volume of diameter d and length ℓ, with its left end a distance ℓ from the dipole center, as shown in Fig. 26-37. Assume that ℓ is much greater than the dipole spacing. *Hint:* Since the cylinder is very thin, you can use the on-axis dipole field (Equation 23-5b) for the field throughout the cylinder.

Solution

The field on the x axis from the dipole is approximately $E_x = 2kp/x^3$, so the energy density is approximately $U = \frac{1}{2}\varepsilon_0E^2 = kp^2/2\pi x^6$. The volume element of the cylinder can be taken to be a disk of area $\pi d^2/4$ and thickness dx. Thus,

$$U = \int u\, dV = \int_\ell^{2\ell}\left(\frac{\pi d^2}{4}\right)\left(\frac{kp^2}{2\pi}\right)\frac{dx}{x^6}$$
$$= \frac{kp^2d^2}{8}\left|-\frac{1}{5x^5}\right|_\ell^{2\ell} = \frac{kp^2d^2}{40\ell^5}\left(1 - \frac{1}{2^5}\right)$$
$$= 31kp^2d^2/1280\ell^5.$$

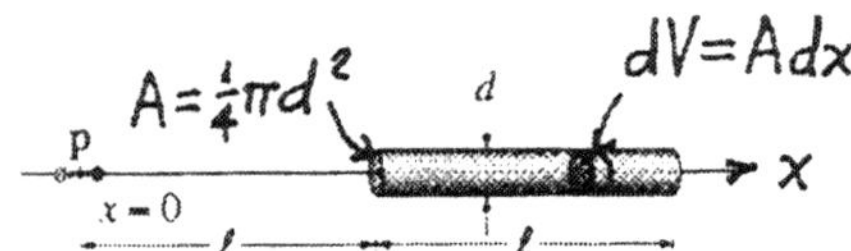

FIGURE 26-37 Problem 87 Solution.

Problem

89. A TV antenna cable consists of two 0.50-mm-diameter wires spaced 12 mm apart. Estimate the capacitance per unit length of this cable, neglecting dielectric effects of the insulation.

Solution

The capacitance per unit length for a bifilar cable, in air, when the diameter of the wires is small compared to their separation, is calculated in the solution to Problem 25-73:

$$\frac{C}{L} = \frac{\lambda}{\Delta V} = \frac{\pi\varepsilon_0}{\ln[(b/a)-1]}$$
$$= \frac{\pi(8.85\ \text{pF/m})}{\ln[(12/0.25)-1]} = 7.22\ \text{pF/m}.$$

Problem

91. Use the fact that the static electric field is conservative to argue that there *must* be fringing field at the edges of a parallel plate capacitor. *Hint:* Remember that the plates are equipotentials, and consider the potential differences V_{AB} and V_{CD} in Fig. 26-38. What does your argument say about the strength of the fringing field relative to the field between the plates?

Solution

The potential difference between the plates, via path $A \rightarrow B$ is $V_{AB} = -\int_{A\rightarrow B} \mathbf{E}\cdot d\boldsymbol{\ell} \neq 0$, since the field is non-zero and parallel to $d\boldsymbol{\ell}$. If there were no fringing field, then the integral of $\mathbf{E}$ over path $C \rightarrow D$ would vanish, $V_{CD} = -\int_{C\rightarrow D} \mathbf{E}\cdot d\boldsymbol{\ell} = 0$, in contradiction to the path-independence of a conservative field. Since the plates are equipotentials, V_{AB} must equal V_{CD}. If we choose a path along an electric field line, we can define an average field strength by $\int \mathbf{E}\cdot d\boldsymbol{\ell} = \int E d\ell = E_{\text{av}}\boldsymbol{\ell}$. It is clear that the average field strength is weaker along the longer field line between the same two points.

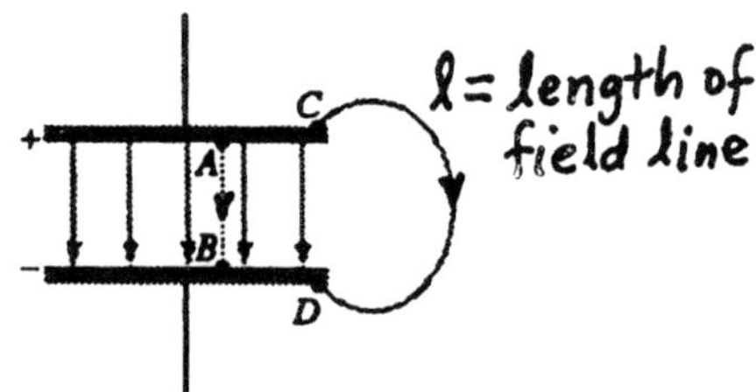

FIGURE 26-38 Problem 91 Solution.

CHAPTER 27 ELECTRIC CURRENT

Section 27-1: Electric Current

Problem

1. A wire carries 1.5 A. How many electrons pass through the wire in each second?

Solution

The current is the amount of charge passing a given point in the wire, per unit time, so in one second, $\Delta q = I\Delta t = (1.5\text{ A})(1\text{ s}) = 1.5\text{ C}$. The number of electrons in this amount of charge is $1.5\text{ C}/1.6\times10^{-19}\text{ C} = 9.38\times10^{18}$.

Problem

3. A 12-V car battery is rated at 80 ampere-hours, meaning it can supply 80 A of current for 1 hour before it becomes discharged. If you accidentally leave the headlights on until the battery discharges, how much charge moves through the lights?

Solution

A battery rated at 80 A·h can supply a net charge of $\Delta q = I\Delta t = (80\text{ C/s})(3600\text{ s}) = 2.88\times10^5\text{ C}$.

Problem

5. Microbiologists measure total current due to potassium ions (K^+) moving through a cell membrane of a rock crab neuron cell to be 30 nA. How many ions pass through the membrane each second?

Solution

The charge moving through the membrane each second is 30 nC. Since singly-charged ions carry one elementary charge (about 160 zC), this corresponds to $30\text{ nC}/(160\text{ zC/ion}) = (1.88\times10^{11}\text{ ion})/(6.02\times10^{23}\text{ ion/mol}) = 0.311\text{ pmol}$. (See Example 27-2(a), Appendix B, and Table 1-1.) Chemical drug-testing instrumentation can detect amounts of substances this low.

Problem

7. The National Electrical Code specifies a maximum current of 10 A in 16-gauge (0.129 cm diameter) copper wire. What is the corresponding current density?

Solution

The cross-section of a wire is uniform, so Equation 27-3a gives $J = I/\frac{1}{4}\pi d^2 = 10\text{ A}/\frac{1}{4}\pi(0.129\text{ cm})^2 = 7.65\text{ MA/m}^2$.

Problem

9. What is the drift speed in a silver wire carrying a current density of 150 A/mm^2? Each silver atom contributes 1.3 free electrons.

Solution

Calculating the density of conduction electrons as in Example 27-1, and using Equation 27-3a for the drift speed, we find $n = (1.3)(10.5\times10^3\text{ kg/m}^3)\div(107.87\text{ u/ion})(1.66\times10^{-27}\text{ kg/u}) = 7.62\times10^{28}\text{ m}^{-3}$, and $v_d = J/ne = (150\text{ A/mm}^2)/(7.62\times10^{28}\text{ m}^{-3})\times(1.6\times10^{-19}\text{ C}) = 1.23\text{ cm/s}$.

Problem

11. A gold film in an integrated circuit measures 2.5 μm thick by 0.18 mm wide. It carries a current density of $6.8\times10^5\text{ A/m}^2$. What is the total current?

Solution

(Assuming the current density is perpendicular to the cross-sectional area of the film) we find, from Equation 27-3a, $I = JA = (6.8\times10^5\text{ A/m}^2)(2.5\ \mu\text{m}\times0.18\text{ mm}) = 306\ \mu\text{A}$.

Problem

13. A plasma used in fusion research contains 5.0×10^{18} electrons and an equal number of protons per cubic meter. Under the influence of an electric field the electrons drift in one direction at 40 m/s, while the protons drift in the opposite direction at 6.5 m/s. (a) What is the current density? (b) What fraction of the current is carried by the electrons?

Solution

The proton current density is $J_p = nev_{d,p} = (5.0\times10^{18}\text{ m}^{-3})(1.6\times10^{-19}\text{ C})(6.5\text{ m/s}) = 5.20\text{ A/m}^2$ (positive in the direction of $v_{d,p}$), and the electron current density is $J_e = n(-e)v_{d,e} = (5.0\times10^{18}\text{ m}^{-3})\times$

$(-1.6\times10^{-19}\ \text{C})(-40\ \text{m/s}) = 32.0\ \text{A/m}^2$. (a) The total current density is $J_p + J_e = 37.2\ \text{A/m}^2$, and (b) the fraction carried by electrons is $32.0/37.2 = 86.0\%$.

Problem

15. In a study of proteins mediating cell membrane transport, microbiologists measure current versus time through the cell membranes of oocytes (nearly mature egg cells) taken from the African clawed frog, *Xenopus*. The measured current versus time is given approximately by $I = 60t + 200t^2 + 4.0t^3$, with t in seconds and I in nA. Find the total charge that flows through the cell membrane in the interval from $t = 0$ to $t = 5.0$ s.

Solution

(a) If we use Equation 27-1b for the charge, then $q = \int_0^{5\,\text{s}} I\,dt = \Big|(60\ \text{nA/s})\frac{1}{2}t^2 + (200\ \text{nA/s}^2)\frac{1}{3}t^3 + (4\ \text{nA/s}^3)\frac{1}{4}t^4\Big|_0^{5\,\text{s}} = (30 + \frac{1}{3}\times1000 + 25)(25\ \text{nC}) = 9.71\ \mu\text{C}$.

Section 27-2: Conduction Mechanisms

Problem

17. What electric field is necessary to drive a 7.5-A current through a silver wire 0.95 mm in diameter?

Solution

From Ohm's law (which applies to silver) and the definition of current density (which we assume is uniform in the wire) one finds $E = \rho J = \rho I/\frac{1}{4}\pi d^2 = (1.59\times10^{-8}\ \Omega\cdot\text{m})(7.5\ \text{A})/\frac{1}{4}\pi(0.95\ \text{mm})^2 = 0.168\ \text{V/m}$.

Problem

19. A 1.0-cm-diameter rod carries a 50-A current when the electric field in the rod is 1.4 V/m. What is the resistivity of the rod material?

Solution

If the rod has a uniform current density and obey's Ohm's law (Equations 27-3a and 4b), then its resistivity is $\rho = E/J = E/(I/\frac{1}{4}\pi d^2) = \frac{1}{4}\pi(10^{-2}\ \text{m})^2\times(1.4\ \text{V/m})/(50\ \text{A}) = 2.20\times10^{-6}\ \Omega\cdot\text{m}$.

Problem

21. Use Table 27-1 to determine the conductivity of (a) copper and (b) sea water.

Solution

Equations 27-4a and b show that the conductivity and the resistivity are reciprocals of one another. Thus, (a) $\rho^{-1} = \sigma = (1.68\times10^{-8}\ \Omega\cdot\text{m})^{-1} = 5.95\times10^7(\Omega\cdot\text{m})^{-1}$ for copper, and (b) $\sigma = (0.22\ \Omega\cdot\text{m})^{-1} = 4.55(\Omega\cdot\text{m})^{-1}$ for typical seawater. (The salinity of open-ocean water varies between 33 and 37‰, but can vary from 1 to 80‰ in shallow coastal waters.)

Problem

23. The free-electron density in aluminum is $2.1\times10^{29}\ \text{m}^{-3}$. What is the collision time in aluminum?

Solution

We can estimate τ from Equation 27-5 and Table 27-1: $\tau = m_e/\rho n e^2 = (9.11\times10^{-31}\ \text{kg})(2.65\times10^{-8}\ \Omega\cdot\text{m})^{-1}\times(2.1\times10^{29}\ \text{m}^{-3})^{-1}(1.6\times10^{-19}\ \text{C})^{-2} = 6.39\times10^{-15}$ s. (An explicit confirmation of the units is:

$$\frac{\text{kg}}{(\Omega\cdot\text{m})(\text{m}^{-3})(\text{C}^2)} = \frac{\text{kg}\cdot\text{m}^2}{(\text{V/A})\text{C}^2} = \frac{\text{kg}\cdot\text{m}^2(\text{C/s})}{(\text{J/C})\text{C}^2} = \frac{\text{kg}\cdot\text{m}^2/\text{s}}{\text{kg}\cdot\text{m}^2/\text{s}^2} = \text{s.}$$
)

Problem

25. The resistivity of copper as a function of temperature is given approximately by $\rho = \rho_0[1 + \alpha(T - T_0)]$, where ρ_0 is the value listed in Table 27-1 for 20°C, $T_0 = 20$°C, and $\alpha = 4.3\times10^{-3}\,^\circ\text{C}^{-1}$. Find the temperature at which copper's resistivity is twice its room-temperature value.

Solution

Taking room temperature to be 20°C, we find that the resistivity doubles when $2 = 1 + \alpha(T - T_0)$, or $T = T_0 + 1/\alpha = 20^\circ\text{C} + 1^\circ\text{C}/4.3\times10^{-3} = 253^\circ\text{C}$.

Section 27-3: Resistance and Ohm's Law

Problem

27. What is the resistance of a heating coil that draws 4.8 A when the voltage across it is 120 V?

Solution

The macroscopic form of Ohm's Law is probably applicable to the heating coil, which is typically a coil of wire. Equation 27-6 gives $R = V/I = 120\ \text{V}\div 4.8\ \text{A} = 25\ \Omega$.

Problem

29. What is the current in a 47-kΩ resistor with 110 V across it?

Solution

Provided the resistor obeys Ohm's law, $I = V/R = 110\text{ V}/47\text{ k}\Omega = 2.34\text{ mA}$.

Problem

31. What current flows when a 45-V potential difference is imposed across a 1.8-kΩ resistor?

Solution

If the resistor obeys Ohm's law, $I = V/R = 45\text{ V}/1.8\text{ k}\Omega = 25\text{ mA}$.

Problem

33. The presence of a few ions makes air a conductor, albeit a poor one. If the total resistance between the ionosphere and Earth is 200 Ω, how much current flows as a result of a 300-kV potential difference between Earth and ionosphere?

Solution

If the total atmospheric resistance is 200 Ω at 300 kV, Equation 27-6 gives a current of $I = V/R = 300\text{ kV}/200\ \Omega = 1.5\text{ kA}$.

Problem

35. A cylindrical iron rod measures 88 cm long and 0.25 cm in diameter. (a) Find its resistance. If a 1.5-V potential difference is applied between the ends of the rod, find (b) the current, (c) the current density, and (d) the electric field in the rod.

Solution

(a) Equation 27-7 gives the resistance of a uniform object of Ohmic material, $R = \rho\ell/A = (9.71\times 10^{-8}\ \Omega\cdot\text{m})(88\text{ cm})/\frac{1}{4}\pi(0.25\text{ cm})^2 = 17.4\text{ m}\Omega$ (see Table 27-1 for the resistivity of iron). (b) Equation 27-6 (Ohm's law) gives $I = V/R = 1.5\text{ V}\div 17.4\text{ m}\Omega = 86.2\text{ A}$. (c) Equation 27-3a gives $J = I/\frac{1}{4}\pi d^2 = 86.2\text{ A}/\frac{1}{4}\pi(0.25\text{ cm})^2 = 17.6\text{ MA/m}^2$. (d) Equation 27-4b gives $E = \rho J = (9.71\times 10^{-8}\ \Omega\cdot\text{m})(17.6\text{ MA/m}^2) = 1.70\text{ V/m}$. (The quantities were calculated in the order queried; alternatively, in reverse order, $E = V/\ell$, $J = E/\rho$, $I = JA$, and $R = V/I$.)

Problem

37. How must the diameters of copper and aluminum wire be related if they are to have the same resistance per unit length?

Solution

From Equation 27-7, the resistance per unit length (of a uniform wire of Ohmic material) is $R/\ell = \rho/\frac{1}{4}\pi d^2$, so equal values for copper and aluminum wires imply that $\rho_{\text{Cu}}/d_{\text{Cu}}^2 = \rho_{\text{Al}}/d_{\text{Al}}^2$, or $d_{\text{Al}}/d_{\text{Cu}} = \sqrt{\rho_{\text{Al}}/\rho_{\text{Cu}}} = \sqrt{2.65/1.68} = 1.26$ (see Table 27-1 for the resistivities).

Problem

39. Engineers call for a power line with a resistance per unit length of 50 mΩ/km. What wire diameter is required if the line is made of (a) copper or (b) aluminum? (c) If the costs of copper and aluminum wire are \$1.53/kg and \$1.34/kg, which material is more economical? The densities of copper and aluminum are 8.9 g/cm^3 and 2.7 g/cm^3, respectively.

Solution

From Equation 27-7, $R/\ell = \rho/(\frac{1}{4}\pi d^2)$, so $d = 2\sqrt{\rho/\pi(R/\ell)}$. With resistivities from Table 27-1, (a) $d_{\text{Cu}} = 2\sqrt{(1.68\times 10^{-8}\ \Omega\cdot\text{m})/(50\pi\text{ m}\ell/\text{km})} = 2.07\text{ cm}$, and (b) $d_{\text{Al}} = \sqrt{2.65/1.68}\, d_{\text{Cu}} = 2.60\text{ cm}$ (see Problem 37). (c) With these diameters, the cost of one meter of wire is $\frac{1}{4}\pi d_{\text{Cu}}^2(1\text{ m})(8.9\text{ g/cm}^3)(\$1.53/\text{kg}) = \$4.58$ for copper, and \$1.92 for aluminum.

Problem

41. Corrosion at battery terminals results in increased resistance, and is a frequent cause of hard starting in cars. In an effort to diagnose hard starting, a mechanic measures the voltage between the battery terminal and the wire carrying current to the starter motor. While the motor is cranking, this voltage is 4.2 V. If the motor draws 125 A, what is the resistance at the battery terminal?

Solution

From Ohm's law, $R = V/I = 4.2\text{ V}/125\text{ A} = 33.6\text{ m}\Omega$. (The resistance of the battery cable in Example 27-5 was 0.60 mΩ, so most of the resistance just calculated was in the connection.)

Section 27-4: Electric Power

Problem

43. A car's starter motor draws 125 A with 11 V across its terminals. What is its power consumption?

Solution

Equation 27-8 gives the power supplied to the motor, $\mathcal{P} = VI = (11\text{ V})(125\text{ A}) = 1.38\text{ kW}$.

Problem

45. A watch uses energy at the rate of 240 μW. How much current does it draw from its 1.5-V battery?

Solution

Rearranging Equation 27-8, we find $I = \mathcal{P}/V = 240\ \mu\text{W}/1.5\ \text{V} = 160\ \mu\text{A}$.

Problem

47. What is the resistance of a standard 120-V, 60-W light bulb?

Solution

The bulb's resistance, from Equation 27-9b, is $R = V^2/\mathcal{P} = (120\ \text{V})^2/60\ \text{W} = 240\ \Omega$, at its operating temperature. (This equation, with average values of voltage and power, can be used for an ac-resistor.)

Problem

49. If the electrons of Problem 4 are accelerated through a potential difference of 10 kV, how much power must be supplied to produce the electron beam?

Solution

The electron beam in Problem 4 carries a current of 4.8 mA, and so must be supplied with a power of $\mathcal{P} = VI = (10\ \text{kV})(4.8\ \text{mA}) = 48$ W. (Recall Equation 27-2, $I = (nA)ev_d = (5.0\times10^6\ \text{electrons/cm})\times(1.6\times10^{-19}\ \text{C/electron})(6.0\times10^7\ \text{m/s})$, where nA is the electron density per length of beam in Problem 4.)

Problem

51. How much total energy could the 12-V battery of Problem 3 supply?

Solution

If the battery is a typical 12 V automotive model, it can supply a total energy of $(12\ \text{V})(80\ \text{A})(1\ \text{h}) = (0.960\ \text{kW·h})(3600\ \text{s/h}) = 3.46$ MJ. (Note that the total energy can be thought of as either the power times 1 h of operating time, $\mathcal{P}\ \Delta t = VI\,\Delta t$, or the total charge times the potential difference, $\Delta qV = (I\,\Delta t)V$ as in Problem 3.

Problem

53. Two cylindrical resistors are made from the same material and have the same length. When connected across the same battery, one dissipates twice as much power as the other. How do their diameters compare?

Solution

At the same voltage, the ratio of the power dissipated is the inverse of the ratio of the resistances, which in turn, goes as the inverse of the square of the ratio of the diameters: $\mathcal{P}_1/\mathcal{P}_2 = (V^2/R_1)/(V^2/R_2) = R_2/R_1 = (\rho\ell/\frac{1}{4}\pi d_2^2)/(\rho\ell/\frac{1}{4}\pi d_1^2) = (d_1/d_2)^2$. We used Equation 27-9b for the power, and Equation 27-7 for the resistance. Thus, if $\mathcal{P}_1 = 2\mathcal{P}_2$, then $d_1 = \sqrt{2}\ d_2$.

Problem

55. A 2000-horsepower electric railroad locomotive gets its power from an overhead wire with 0.20 Ω/km. The potential difference between wire and track is 10 kV. Current returns through the track, whose resistance is negligible. (a) How much current does the locomotive draw? (b) How far from the power plant can the train go before 1% of the energy is lost in the wire?

Solution

(a) If we neglect possible energy losses in the locomotive's engine, $\mathcal{P}_{\text{in}} = \mathcal{P}_{\text{out}}$, or $VI = 2000$ hp. Thus, $I = (2000\ \text{hp})(746\ \text{W/hp})/(10^4\ \text{V}) = 149$ A.
(b) The power loss in getting current to and from the locomotive is $\mathcal{P}_{\text{loss}} = I^2R$, where I is the current from part (a) and R is the resistance of the feed/return circuit. The high voltage wire has resistance $(0.2\ \Omega/\text{km})\ell$, where its length, ℓ, is also the distance from the power station, and the resistance of the return rail is negligible (see Problem 30). Then $\mathcal{P}_{\text{loss}} = 1\%, \mathcal{P}_{\text{in}}$ implies $I^2R = 0.01VI$, or $\ell = (0.01)(10^4\ \text{V})/(149\ \text{A})(0.2\ \Omega/\text{km}) = 3.35$ km.

Paired Problems

Problem

57. Electrons in a fine silver wire 20 μm in diameter drift at 0.14 mm/s. What is the current in the wire? Each silver atom contributes 1.3 free electrons.

Solution

For a uniform wire, Equation 27-2 gives $I = ne(\frac{1}{4}\pi d^2)v_d = (7.62\times10^{28}\ \text{m}^{-3})(1.6\times10^{-19}\ \text{C})\times\frac{1}{4}\pi(20\ \mu\text{m})^2(0.14\ \text{mm/s}) = 536\ \mu\text{A}$. We used the density of conduction electrons for silver from Problem 9.

Problem

59. What is the resistance of a column of mercury 0.75 m long and 1.0 mm in diameter?

Solution

At ordinary temperatures, metallic mercury obeys Ohm's law, so Equation 27-7 and Table 27-1 give a resistance of $R = \rho\ell/A = (9.84\times10^{-7}\ \Omega\cdot\text{m})\times(0.75\ \text{m})/\frac{1}{4}\pi(10^{-3}\ \text{m})^2 = 0.940\ \Omega$.

Problem

61. A power plant produces 1000 MW to supply a city 40 km away. Current flows from the power plant on a single wire of resistance 0.050 Ω/km, through the city, and returns via the ground, assumed to have negligible resistance. At the power plant the voltage between the wire and ground is 115 kV. (a) What is the current in the wire? (b) What fraction of the power is lost in transmission?

Solution

(a) If the power supplied by the plant is 1 GW at 115 kV, then the current supplied is $I = \mathcal{P}_{\text{out}}/V = 10^9\ \text{W}/115\ \text{kV} = 8.70\ \text{kA}$. (b) The power loss in the transmission line is $\mathcal{P}_{\text{loss}} = I^2R_{\text{line}} = (8.70\ \text{kA})^2\times(0.050\ \Omega/\text{km}\times 40\ \text{km}) = 151\ \text{MW}$, or 15.1% of the plants 1 GW output. (Note: The voltage drop along the transmission line is $IR_{\text{line}} = 17.4$ kV, not 115 kV.)

Problem

63. A 240-V electric motor is 90% efficient, meaning that 90% of the energy supplied to it ends up as mechanical work. If the motor lifts a 200-N weight at 3.1 m/s, how much current does it draw?

Solution

The electrical power input to the motor is $\mathcal{P}_{\text{in}} = VI$, while the mechanical power output (lifting the weight) is $\mathcal{P}_{\text{out}} = Fv = 90\%\ \mathcal{P}_{\text{in}}$. Thus, $I = Fv/0.9\ V = (200\ \text{N})(3.1\ \text{m/s})/(0.9\times240\ \text{V}) = 2.87\ \text{A}$.

Supplementary Problems

Problem

65. A metal bar has a rectangular cross section 5.0 cm by 10 cm, as shown in Fig. 27-25. The bar has a nonuniform conductivity, ranging from zero at the bottom to a maximum at the top. As a result, the current density increases linearly from zero at the bottom to 0.10 A/cm^2 at the top. What is the total current in the bar?

Solution

The current density is $\mathbf{J} = (0.1\ \text{A/cm}^2)(x/10\ \text{cm})\hat{\mathbf{k}}$, in the coordinate system superposed on Fig. 27-25. The cross section can be divided into strips of area $d\mathbf{A} = (5\ \text{cm})\,dx\,\hat{\mathbf{k}}$ (over which $\mathbf{J}$ is constant), so the total current in the bar (Equation 27-3a in differential form) is:

$$I = \int_{x\text{-sect}} \mathbf{J}\cdot d\mathbf{A} = \int_0^{10\ \text{cm}} (10^{-2}\ \text{A/cm}^3)(5\ \text{cm})x\,dx$$
$$= (0.05\ \text{A/cm}^2)\frac{1}{2}(10\ \text{cm})^2 = 2.5\ \text{A}.$$

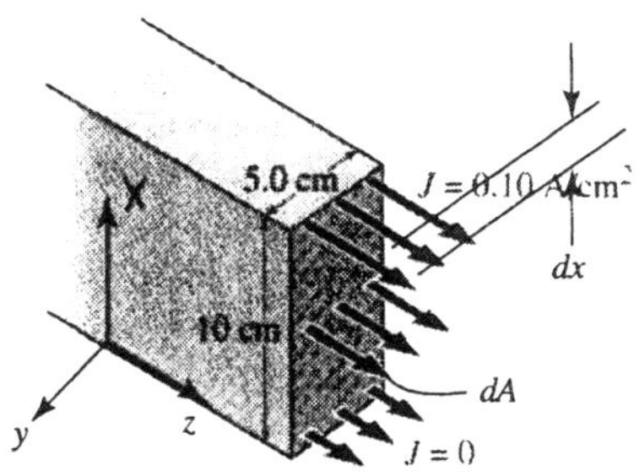

FIGURE 27-25 Problem 65 Solution.

Problem

67. General Motors' EV1 electric car has a mass of 1500 kg and is powered by 26 12-V batteries connected in series, for a total of 312 V. About 85% of the electrical energy from the batteries ends up as mechanical energy at the drive wheels. How much current do the batteries supply when the car is climbing a 10° slope at 45 km/h? Neglect frictional losses and air resistance.

Solution

If frictional losses and air resistance are neglected, the motor only supplies power to work against gravity, $\mathcal{P}_{\text{out}} = F_{\|}v = (mg\sin 10^\circ)v$, which equals 85% of the input electrical power from the batteries, $\mathcal{P}_{\text{in}} = VI$. Therefore, $I = mg\sin 10^\circ v/0.85V = (1500\times9.8\ \text{N})\times(45\ \text{m}/3.6\ \text{s})\sin 10^\circ/(0.85\times312\ \text{V}) = 120\ \text{A}$.

Problem

69. A 100-Ω resistor of negligible mass is mounted inside a calorimeter. When a 12-V battery is connected for 5.0 min, the temperature inside the calorimeter rises by 26°C. What is the heat capacity of the calorimeter contents?

Solution

The electrical energy dissipated in the resistor raises the temperature inside the calorimeter, $\mathcal{P}t = (V^2/R)t = C\,\Delta T$, where C is the heat capacity of the contents (the heat capacity of a resistor of negligible mass is approximately zero). Therefore, $C =$

$V^2t/R\,\Delta T = (12\text{ V})^2(5\times 60\text{ s})/(100\ \Omega)(26\text{ C}^\circ) = 16.6\text{ J/C}^\circ$.

Problem

71. Figure 27-27 shows a resistor made from a truncated cone of material with uniform resistivity ρ. Consider the cone to be made of thin slices of thickness dx, like the one shown; Equation 27-7 shows that the resistance of each slab is $dR = \rho\,dx/A$. By integrating over all such slices, shows that the resistance between the two flat faces is $R = \rho\ell/\pi ab$. (This method assumes the equipotentials are planes, which is only approximately true.)

Solution

As suggested, consider that the total resistance R equals $\int_{\text{cone}} dR$, where $dR = \rho\,dx/A$ is the resistance of a thin disk as shown on Fig. 27-27. The area of such a disk is $A = \pi y^2$, where $y = a + (b-a)x/\ell$ is the radius and $0 \le x \le \ell$ as shown. Then

$$R = \int_0^\ell \frac{\rho\,dx}{\pi[a+(b-a)x/\ell]^2} = \frac{\rho}{\pi}\left|-\left(\frac{\ell}{b-a}\right)\frac{1}{[a+(b-a)x/\ell]}\right|_0^\ell = \frac{\rho}{\pi}\left(\frac{\ell}{b-a}\right)\left(\frac{1}{a}-\frac{1}{b}\right) = \frac{\rho\ell}{\pi ab}.$$

(This result depends on the condition that the flat faces and parallel circular cross-sections of the cone are equipotential surfaces.)

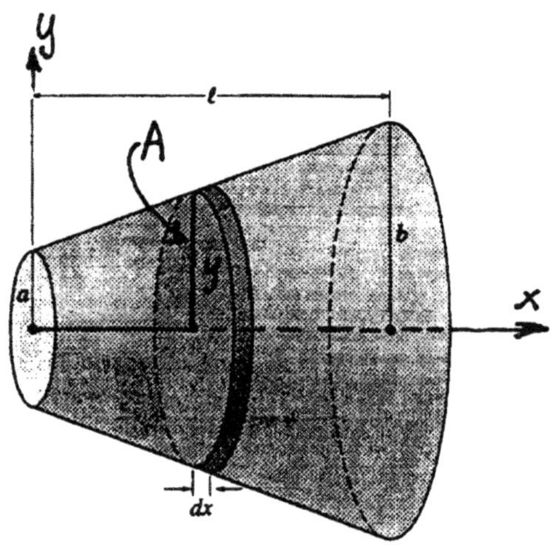

FIGURE 27-27 Problem 71 Solution.

Problem

73. At some point in a material of resistivity ρ the current density is J. Show that the power per unit volume dissipated at that point is $J^2\rho$.

Solution

Consider a small volume in the material, ΔV, where the current density is $\mathbf{J} = nq\,\mathbf{v}_d$ (for simplicity, we assume only one type of charge carrier). The electric field does work on each charge carrier at a rate equal to $\mathbf{F}_{\text{el}}\cdot\mathbf{v}_d = q\,\mathbf{E}\cdot\mathbf{v}_d$, so the total power is this times the number of charge carriers in ΔV, or $\Delta\mathcal{P} = (n\,\Delta V)(q\,\mathbf{E}\cdot\mathbf{v}_d) = \mathbf{J}\cdot\mathbf{E}\,\Delta V$. If the electric field in the material is given by Ohm's law (such a material is called a passive region; if other effective fields are present, the material is an active region), then $\mathbf{E} = \rho\mathbf{J}$, where ρ is the resistivity, and $\Delta\mathcal{P}/\Delta V = \mathbf{J}\cdot(\rho\mathbf{J}) = \rho J^2$. This power is transformed into random thermal motion of the charge carriers and ions of the material, and is referred to as Joule heat; $J^2\rho$ is the rate of Joule heating per unit volume of material.

CHAPTER 28 ELECTRIC CIRCUITS

ActivPhysics can help with these problems: Section 12, "DC Circuits"

Section 28-1: Circuits and Symbols

Problem

1. Sketch a circuit diagram for a circuit that includes a resistor R_1 connected to the positive terminal of a battery, a pair of parallel resistors R_2 and R_3 connected to the lower-voltage end of R_1, then returned to the battery's negative terminal, and a capacitor across R_2.

Solution

A literal reading of the circuit specifications results in connections like those in sketch (a). Because the connecting wires are assumed to have no resistance (a real wire is represented by a separate resistor), a topologically equivalent circuit diagram is shown in sketch (b).

Problem 1 Solution (a).

Problem 1 Solution (b).

Problem

3. Resistors R_1 and R_2 are connected in series, and this series combination is in parallel with R_3. This parallel combination is connected across a battery whose internal resistance is R_{int}. Draw a diagram representing this circuit.

Solution

The circuit has three parallel branches: one with R_1 and R_2 in series; one with just R_3; and one with the battery (an ideal emf in series with the internal resistance).

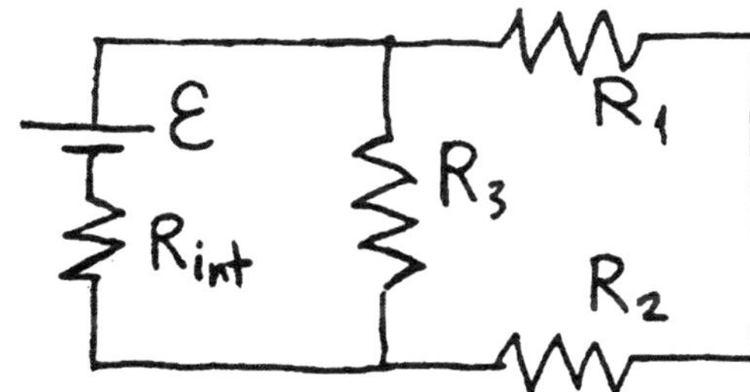

Problem 3 Solution.

Section 28-2: Electromotive Force

Problem

5. A 1.5-V battery stores 4.5 kJ of energy. How long can it light a flashlight bulb that draws 0.60 A?

Solution

The average power, supplied by the battery to the bulb, multiplied by the time equals the energy capacity of the battery. For an ideal battery, $\mathcal{P} = \mathcal{E}I$, therefore $\mathcal{E}It = 4.5$ kJ, or $t = 4.5\text{ kJ}/(1.5\text{ V})\times(0.60\text{ A}) = 5\times10^3\text{ s} = 1.39\text{ h}$.

Problem

7. A battery stores 50 W·h of chemical energy. If it uses up this energy moving 3.0×10^4 C through a circuit, what is its voltage?

Solution

The emf is the energy (work done going through the source from the negative to the positive terminal) per unit charge: $\mathcal{E} = (50\text{ W·h})(3600\text{ s/h})/(3\times10^4\text{ C}) = 6\text{ V}$. (This is the average emf; the actual emf may vary with time.)

Section 28-3: Simple Circuits: Series and Parallel Resistors

Problem

9. What resistance should be placed in parallel with a 56-kΩ resistor to make an equivalent resistance of 45 kΩ?

Solution

The solution for R_2 in Equation 28-3a is $R_2 = R_1\ R_{\text{parallel}}/(R_1 - R_{\text{parallel}}) = (56\ \text{k}\Omega)(45)/(56-45) = 229\ \text{k}\Omega$.

Problem

11. In Fig. 28-49, take all resistors to be 1.0 Ω. If a 6.0-V battery is connected between points A and B, what will be the current in the vertical resistor?

Solution

The circuit in Fig. 28-49, with a battery connected across points A and B, is similar to the circuit analysed in Example 28-4. In this case, $R_{||} = (1\ \Omega)\times(2)/(1+2) = (\frac{2}{3})\ \Omega$, and $R_{\text{tot}} = 1\ \Omega + 1\ \Omega + \frac{2}{3}\ \Omega = \frac{8}{3}\ \Omega$. The total current (that through the battery) is $I_{\text{tot}} = \mathcal{E}/R_{\text{tot}} = 6\ \text{V}/(\frac{8}{3}\ \Omega) = (\frac{9}{4})\ \text{A}$. The voltage across the parallel combination is $I_{\text{tot}}R_{||} = (\frac{9}{4}\ \text{A})\times(\frac{2}{3}\ \Omega) = \frac{3}{2}\ \text{V}$, which is the voltage across the vertical 1 Ω resistor. The current through this resistor is then $(\frac{3}{2}\ \text{V})/(1\ \Omega) = 1.5\ \text{A}$.

Problem

13. What is the internal resistance of the battery in the preceding problem?

Solution

The solution of the preceding problem (or the reasoning of Example 28-2) gives $R_{\text{int}} = 0.02\ \Omega$ (i.e., $(12\ \text{V} - 6\ \text{V})/300\ \text{A}$).

Problem

15. When a 9-V battery is temporarily short-circuited, a 200-mA current flows. What is the internal resistance of the battery?

Solution

From the equation for a battery short-circuited, in the subsection "Real Batteries," $R_{\text{int}} = \mathcal{E}/I = 9\ \text{V}/0.2\ \text{A} = 45\ \Omega$.

Problem

17. A partially discharged car battery can be modeled as a 9-V emf in series with an internal resistance of 0.08 Ω. Jumper cables are used to connect this battery to a fully charged battery, modeled as a 12-V emf in series with a 0.02-Ω internal resistance. How much current flows through the discharged battery?

Solution

Terminals of like polarity are connected with jumpers of negligible resistance. Kirchhoff's voltage law gives $\mathcal{E}_1 - \mathcal{E}_2 - IR_1 - IR_2 = 0$, or $I = (\mathcal{E}_1 - \mathcal{E}_2)/(R_1 + R_2) = (12-9)\ \text{V}/(0.02+0.08)\ \Omega = 30\ \text{A}$.

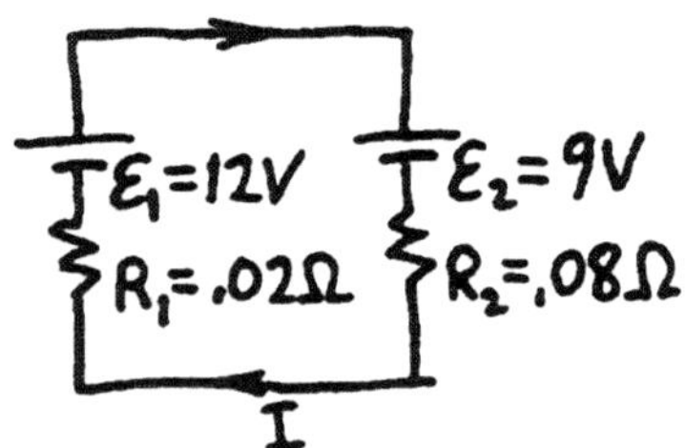

Problem 17 Solution.

Problem

19. What is the equivalent resistance between A and B in each of the circuits shown in Fig. 28-50? *Hint:* In (*c*), think about symmetry and the current that would flow through R_2.

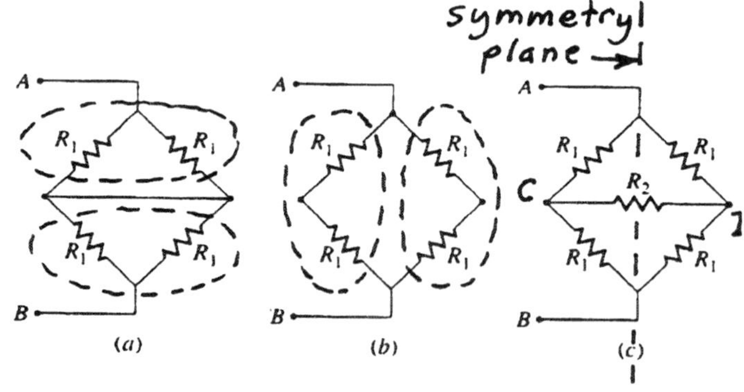

FIGURE 28-50 Problem 19 Solution.

Solution

(a) There are two parallel pairs $(\frac{1}{2}R_1)$ in series, so $R_{AB} = \frac{1}{2}R_1 + \frac{1}{2}R_1 = R_1$. (b) Here, there are two series pairs $(2R_1)$ in parallel, so $R_{AB} = (2R_1)(2R_1)\div(2R_1 + 2R_1) = R_1$. (c) Symmetry requires that the current divides equally on the right and left sides, so points C and D are at the same potential. Thus, no current flows through R_2, and the circuit is equivalent to (b). (Note that the reasoning in parts (a) and (b) is easily generalized to resistances of different values; the generalization in part (c) requires the equality of ratios of resistances which are mirror images in the plane of symmetry.)

Problem

21. How many 100-W, 120-V light bulbs can be connected in parallel before they below a 20-A circuit breaker?

Solution

The circuit breaker is activated if $I = 120\text{ V}\div R_{min} > 20\text{ A}$, or if $R_{min} < 6\ \Omega$. The resistance of each light bulb is $R = V^2/\mathcal{P} = (120\text{ V})^2/100\text{ W} = 144\ \Omega$, and n bulbs in parallel have resistance $R_{||} = R/n$. Therefore $R_{||} \geq R_{min}$ implies $n \leq 144/6 = 24$, so more than 24 bulbs would blow the circuit.

Problem

23. Take $\mathcal{E} = 12$ V and $R_1 = 270\ \Omega$ in the voltage divider of Fig. 28-5. (a) What should be the value of R_2 in order that 4.5 V appear across R_2? (b) What will be the power dissipation in R_2?

Solution

(a) For this voltage divider, Equation 28-2b gives $V_2 = R_2\mathcal{E}/(R_1 + R_2)$, or $R_2 = R_1V_2/(\mathcal{E} - V_2) = (270\ \Omega)(4.5)/(12 - 4.5) = 162\ \Omega$. (b) The power dissipated (Equation 27-9b) is $\mathcal{P}_2 = V_2^2/R_2 = (4.5\text{ V})^2/162\ \Omega = 125$ mW.

Problem

25. In the circuit of Fig. 28-52, R_1 is a variable resistor, and the other two resistors have equal resistances R. (a) Find an expression for the voltage across R_1, and (b) sketch a graph of this quantity as a function of R_1 as R_1 varies from 0 to $10R$. (c) What is the limiting value as $R_1 \to \infty$?

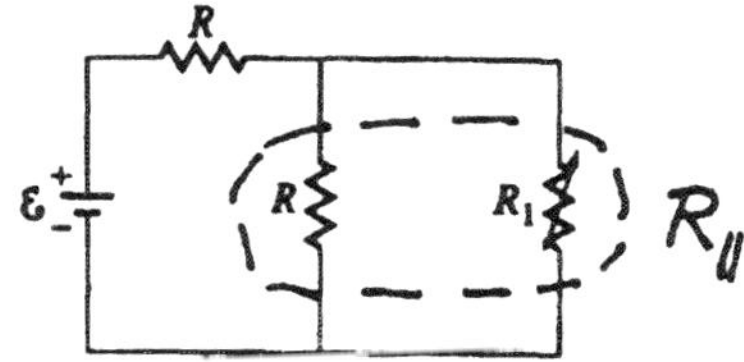

FIGURE 28-52 Problem 25.

Solution

(a) The resistors in parallel have an equivalent resistance of $R_{||} = RR_1/(R + R_1)$. The other R, and $R_{||}$, is a voltage divider in series with $\mathcal{E}$, so Equation 28-2 gives $V_{||} = \mathcal{E}R_{||}/(R + R_{||}) = \mathcal{E}R_1/(R + 2R_1)$. (b) and (c) If $R_1 = 0$ (the second resistor shorted out), $V_{||} = 0$, while if $R_1 = \infty$ (open circuit), $V_{||} = \frac{1}{2}\mathcal{E}$ (the value when R_1 is removed). If $R_1 = 10R$, $V_{||} = (10/21)\mathcal{E}$ (as in Problem 24).

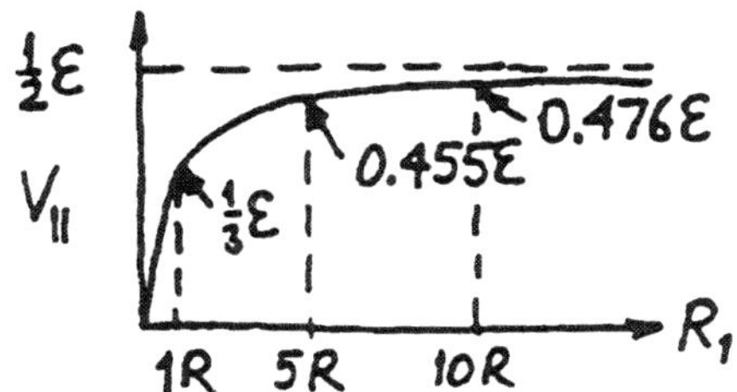

Problem 25 Solution.

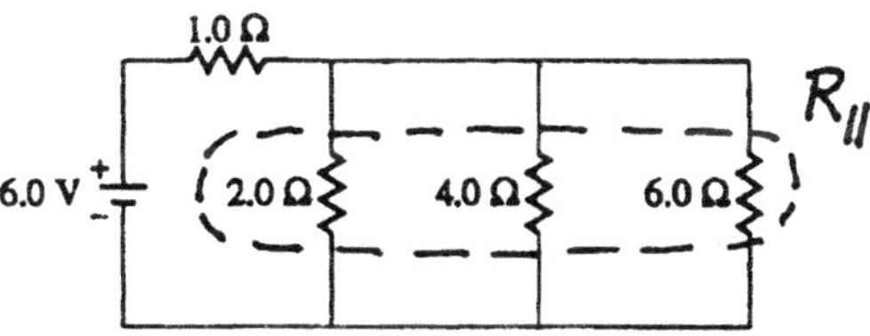

FIGURE 28-53 Problem 26 Solution, and Problem 27.

Problem

27. In the circuit of Fig. 28-53, how much power is being dissipated in the 4-Ω resistor?

Solution

The three resistors in parallel have an effective resistance of $1/R_{||} = (1/2 + 1/4 + 1/6)\ \Omega^{-1}$, or $R_{||} = (12/11)\ \Omega$. Equation 28-2 gives the voltage across them as $V_{||} = (6\text{ V})(12/11)/(1 + 12/11) = (72/23)$ V. Thus, $\mathcal{P}_4 = V_{||}^2/R_4 = (72/23)^2\text{ V}^2/4\ \Omega = 2.45$ W.

Section 28-4: Kirchhoff's Laws and Multiloop Circuits

Problem

29. In the circuit of Fig. 28-54 it makes no difference whether the switch is open or closed. What is $\mathcal{E}_3$ in terms of the other quantities shown?

Solution

If the switch is irrelevant, then there is no current through its branch of the circuit. Thus, points A and B must be at the same potential, and the same current flows through R_1 and R_2. Kirchhoff's voltage law applied to the outer loop, and to the left-hand loop, gives $\mathcal{E}_1 - IR_1 - IR_2 + \mathcal{E}_2 = 0$, and $\mathcal{E}_1 - IR_1 + \mathcal{E}_3 = 0$, respectively. Therefore,

$$\mathcal{E}_3 = IR_1 - \mathcal{E}_1 = \left(\frac{\mathcal{E}_1 + \mathcal{E}_2}{R_1 + R_2}\right)R_1 - \mathcal{E}_1 = \frac{\mathcal{E}_2 R_1 - \mathcal{E}_1 R_2}{R_1 + R_2}.$$

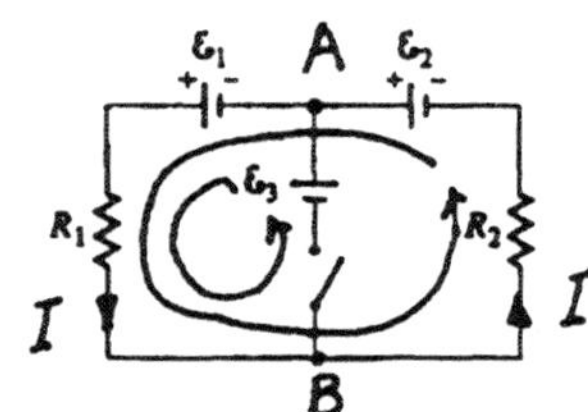

FIGURE 28-54 Problem 29 Solution.

Problem

31. In Fig. 28-56, what is the equivalent resistance measured between points A and B?

Solution

The effective resistance is determined by the current which would flow through a pure emf if it were connected between A and B: $R_{AB} = \mathcal{E}/I$. Since I is but one of six branch currents, the direct solution of Kirchhoff's circuit laws is tedious (6×6 determinants). (The method of loop currents, not mentioned in the text, involves more tractable 3×3 determinants.) However, because of the special values of the resistors in Fig. 28-56, a symmetry argument greatly simplifies the calculation.

The equality of the resistors on opposite sides of the square implies that the potential difference between A and C equals that between D and B, i.e., $V_A - V_C = V_D - V_B$. Equivalently, $V_A - V_D = V_C - V_B$. Since $V_A - V_C = I_1R$, $V_A - V_D = I_2(2R)$, etc., the symmetry argument requires that both R-resistors on the perimeter carry the same current, I_1, and both $2R$-resistors carry current I_2. Then Kirchhoff's current law implies that the current through $\mathcal{E}$ is $I_1 + I_2$, and the current through the central resistor is $I_1 - I_2$ (as added to Fig. 28-56). Now there are only two independent branch currents, which can be found from Kirchhoff's voltage law, applied, for example, to loops $ACBA$, $\mathcal{E} - I_1R - I_2(2R) = 0$, and $ACDA$, $-I_1R - (I_1 - I_2)R + I_2(2R) = 0$. These equations may be rewritten as $I_1 + 2I_2 = \mathcal{E}/R$ and $-2I_1 + 3I_2 = 0$, with solution $I_1 = 3\mathcal{E}/7R$ and $I_2 = 2\mathcal{E}/7R$. Therefore, $I = I_1 + I_2 = 5\mathcal{E}/7R$, and $R_{AB} = \mathcal{E}/I = 7R/5$. (The configuration of resistors in Fig. 28-56 is called a Wheatstone bridge.)

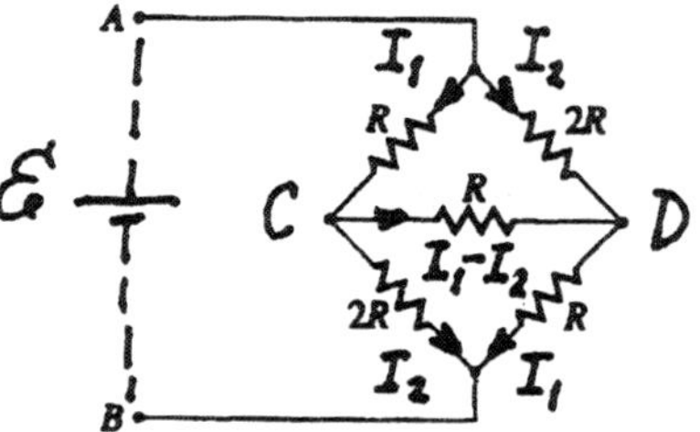

FIGURE 28-56 Problem 31 Solution.

Problem

33. Find all three currents in the circuit of Fig. 28-18 with the values given, but with battery $\mathcal{E}_2$ reversed.

Solution

The general solution of the two loop equations and one node equation given in Example 28-5 can be found using determinants (or I_1 and I_2 can be found in terms of I_3, as in Example 28-5). The equations and the solution are:

$$\begin{aligned} I_1R_1 + 0 + I_3R_3 &= \mathcal{E}_1 \quad &(\text{loop } 1),\\ 0 - I_2R_2 + I_3R_3 &= \mathcal{E}_2 \quad &(\text{loop } 2),\\ I_1 - I_2 - I_3 &= 0 \quad &(\text{node A});\end{aligned}$$

$$\Delta \equiv \begin{vmatrix} R_1 & 0 & R_3 \\ 0 & -R_2 & R_3 \\ 1 & -1 & -1 \end{vmatrix} = R_1R_2 + R_2R_3 + R_3R_1,$$

$$I_1 = \frac{1}{\Delta}\begin{vmatrix} \mathcal{E}_1 & 0 & R_3 \\ \mathcal{E}_2 & -R_2 & R_3 \\ 0 & -1 & -1 \end{vmatrix} = \frac{\mathcal{E}_1(R_2 + R_3) - \mathcal{E}_2R_3}{\Delta},$$

$$I_2 = \frac{1}{\Delta}\begin{vmatrix} R_1 & \mathcal{E}_1 & R_3 \\ 0 & \mathcal{E}_2 & R_3 \\ 1 & 0 & -1 \end{vmatrix} = \frac{\mathcal{E}_1R_3 - \mathcal{E}_2(R_1 + R_3)}{\Delta},$$

$$I_3 = \frac{1}{\Delta}\begin{vmatrix} R_1 & 0 & \mathcal{E}_1 \\ 0 & -R_2 & \mathcal{E}_2 \\ 1 & -1 & 0 \end{vmatrix} = \frac{\mathcal{E}_2R_1 + \mathcal{E}_1R_2}{\Delta}.$$

With the particular values of emf's and resistors in this problem, we find currents of $I_1 = [(4+1)6 - 1(-9)]\ \text{A}/14 = 2.79\ \text{A}$, $I_2 = [1\times 6 - (2+1)(-9)]\ \text{A}/14 = 2.36\ \text{A}$, and $I_3 = [4\times 6 + 2\times (-9)]\ \text{A}/14 = 0.429\ \text{A}$. Or, one could retrace the reasoning of Example 28-5, with $\mathcal{E}_2 = -9\ \text{V}$ replacing the original value in loop 2. Then, everything is the same until the equation $-9 + 2(6 - 3I_3) - I_3 = 0$, or $I_3 = (\frac{3}{7})\ \text{A}$, $I_2 = \frac{1}{2}(6 - 3\times\frac{3}{7})\ \text{A} = (33/14)\ \text{A}$, and $I_1 = I_2 + I_3 = (39/14)\ \text{A}$.

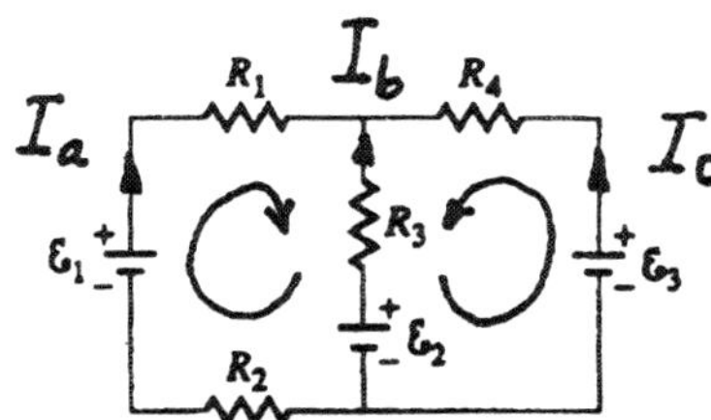

FIGURE 28-57 Problem 34 Solution, and Problems 35 and 36.

Problem

35. With all the values except $\mathcal{E}_2$ in Fig. 28-57 as given in the preceding problem, find the condition on $\mathcal{E}_2$ that will make the current in R_3 flow upward.

Solution

Let us choose the positive sense for each of the three branch currents in Fig. 28-57 as upward through their respective emf's (at least one must be negative, of course), and consider the two smaller loops shown. Kirchhoff's circuit laws give:

$$I_a + I_b + I_c = 0 \qquad \text{(top node)}$$
$$(R_1 + R_2)I_a - R_3 I_b = \mathcal{E}_1 - \mathcal{E}_2 \qquad \text{(left loop)}$$
$$-R_3 I_b + R_4 I_c = \mathcal{E}_3 - \mathcal{E}_2 \qquad \text{(right loop)}.$$

Solve for I_a and I_c from the loop equations and substitute into the node equation:

$$\frac{(\mathcal{E}_1 - \mathcal{E}_2) + R_3 I_b}{R_1 + R_2} + I_b + \frac{(\mathcal{E}_3 - \mathcal{E}_2) + R_3 I_b}{R_4} = 0.$$

Then

$$I_b = \frac{R_4(\mathcal{E}_2 - \mathcal{E}_1) + (R_1 + R_2)(\mathcal{E}_2 - \mathcal{E}_3)}{R_3 R_4 + (R_1 + R_2)(R_3 + R_4)},$$

with similar expressions for I_a and I_c. One can see that I_b is positive if $R_4(\mathcal{E}_2 - \mathcal{E}_1) + (R_1 + R_2)\times(\mathcal{E}_2 - \mathcal{E}_3) > 0$, or

$$\mathcal{E}_2 > \frac{R_4\mathcal{E}_1 + (R_1 + R_2)\mathcal{E}_3}{R_1 + R_2 + R_4} = \frac{(820\ \Omega)(6\ \text{V}) + (420\ \Omega)(4.5\ \text{V})}{(1240\ \Omega)} = 5.49\ \text{V}.$$

Problem

37. Figure 28-58 shows a portion of a circuit used to model the electrical behavior of long, cylindrical biological cells such as muscle cells or the axons of neurons. Find the current through the emf $\mathcal{E}_3$, given that all resistors have the same value $R = 1.5\ \text{M}\Omega$ and that $\mathcal{E}_1 = 75$ mV, $\mathcal{E}_2 = 45$ mV, and $\mathcal{E}_3 = 20$ mV. Be sure to specify the direction of the current.

Solution

The circuits in Figs. 28-57 and 58 each have three branches (with emf's and total resistances $\mathcal{E}_a, R_a, \mathcal{E}_b, R_b$, and $\mathcal{E}_c, R_c$) connected in parallel between two nodes. (The emf's and currents in each branch are taken as positive upward in the figures.) The expression for I_b in the solution to Problem 35 can be used for any permutation of indices a, b, and c (any order of parallel branches is equivalent). For example, in Fig. 28-58, take b to be the third branch etc., so that $\mathcal{E}_b = \mathcal{E}_3 = 20$ mV, $R_b = 2R$, $\mathcal{E}_a = 75$ mV, $R_a = 2R$, $\mathcal{E}_c = 45$ mV, and $R_c = R$, with $R = 1.5\ \text{M}\Omega$. Then

$$I_b = \frac{(\mathcal{E}_b - \mathcal{E}_a)R_c + (\mathcal{E}_b - \mathcal{E}_c)R_a}{R_aR_b + R_bR_c + R_cR_a} = \frac{(\mathcal{E}_b - \mathcal{E}_a)R + (\mathcal{E}_b - \mathcal{E}_c)2R}{4R^2 + 2R^2 + 2R^2}$$
$$= (3\mathcal{E}_b - 2\mathcal{E}_c - \mathcal{E}_a)/8R = (60 - 90 - 75)\ \text{mV}/12\ \text{M}\Omega$$
$$= -8.75\ \text{nA}.$$

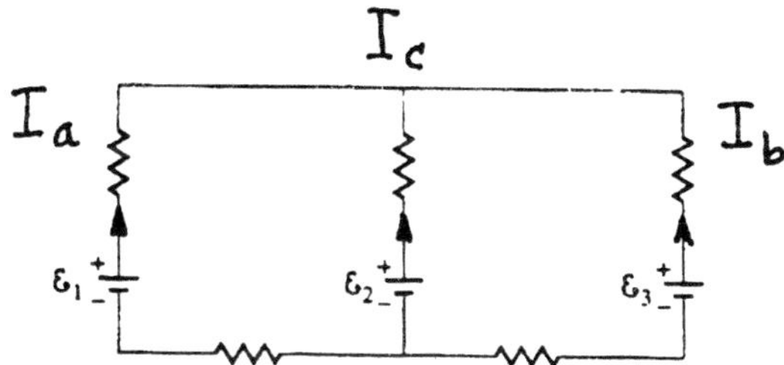

FIGURE 28-58 Problems 37, 38.

Section 28-5: Electrical Measuring Instruments

Problem

39. A voltmeter with 200-kΩ resistance is used to measure the voltage across the 10-kΩ resistor in Fig. 28-59. By what percentage is the measurement in error because of the finite meter resistance?

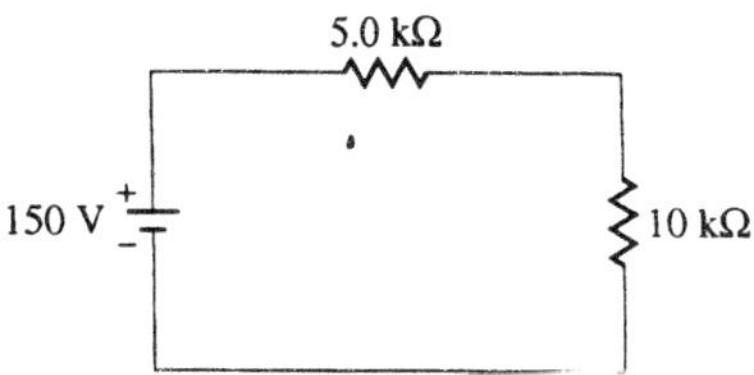

FIGURE 28-59 Problems 39 and 40.

Solution

The voltage across the 10 kΩ resistor in Fig. 28-59 is (150 V)(10)/(10 + 5) = 100 V (the circuit is just a voltage divider as described by Equations 28-2a and b), as would be measured by an ideal voltmeter with infinite resistance. With the real voltmeter connected in parallel across the 10 kΩ resistor, its effective resistance is changed to $R_{\parallel} = (10\ \text{k}\Omega)(200\ \text{k}\Omega) \div (210\ \text{k}\Omega) = 9.52\ \text{k}\Omega$, and the voltage reading is only $(150\ \text{V})(9.52)/(9.52+5) = 98.4\ \text{V}$, or about 1.64% lower.

Problem

41. A neophyte mechanic foolishly connects an ammeter with 0.1-Ω resistance directly across a 12-V car battery whose internal resistance is 0.01 Ω. What is the power dissipation in the meter? No wonder it gets destroyed!

Solution

The current through the misconnected ammeter is $I = \mathcal{E}/(R_{\text{int}} + R_m)$, so the power dissipated in it is $\mathcal{P} = I^2 R_m = \mathcal{E}^2 R_m/(R_{\text{int}} + R_m)^2 = (12\ \text{V}/0.11\ \Omega)^2(0.1\ \Omega) = 1.19\ \text{kW}$ (comparable to a small toaster-oven).

Problem

43. In Fig. 28-61 what are the meter readings when (a) an ideal voltmeter or (b) an ideal ammeter is connected between points A and B?

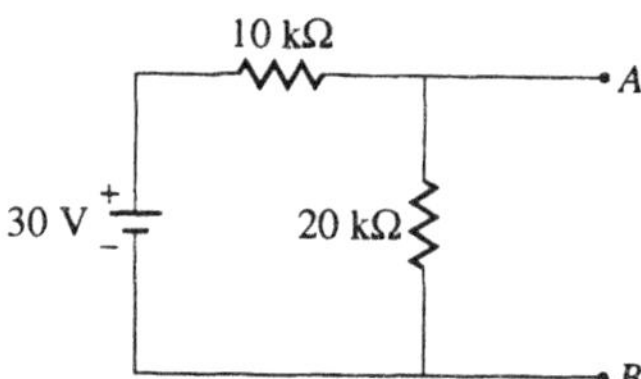

FIGURE 28-61 Problem 43.

Solution

(a) An ideal voltmeter has infinite resistance, so AB is still an open circuit (as shown on Fig. 28-61) when such a voltmeter is connected. The meter reads the voltage across the 20 kΩ resistor (part of a voltage divider), or $(30\ \text{V})20/(20+10) = 20\ \text{V}$ (see Equation 28-2a or b). (b) An ideal ammeter has zero resistance, and thus measures the current through the points A and B when short-circuited (i.e., no current flows through the 20 kΩ resistor). In Fig. 28-61, this would be $I_{AB} = 30\ \text{V}/10\ \Omega = 3\ \text{mA}$. (Such a connection does not measure the current in the original circuit, since an ammeter should be connected in series with the current to be measured.)

Section 28-6: Circuits with Capacitors

Problem

45. Show that the quantity RC has the units of time (seconds).

Solution

The SI units for the time constant, RC, are $(\Omega)(\text{F}) = (\text{V/A})(\text{C/V}) = (\text{s/C})(\text{C}) = \text{s}$, as stated.

Problem

47. Show that a capacitor is charged to approximately 99% of the applied voltage in five time constants.

Solution

After five time constants, Equation 28-6 gives a voltage of $V_C/\mathcal{E} = 1 - e^{-5} = 1 - 6.74\times10^{-3} \simeq 99.3\%$ of the applied voltage.

Problem

49. Figure 28-62 shows the voltage across a capacitor that is charging through a 4700-Ω resistor in the circuit of Fig. 28-29. Use the graph to determine (a) the battery voltage, (b) the time constant, and (c) the capacitance.

Solution

(a) For the circuit considered, the voltage across the capacitor asymptotically approaches the battery voltage after a long time (compared to the time constant). In Fig. 28-62, this is about 9 V. (b) The time constant is the time it takes the capacitor voltage to reach $1 - e^{-1} = 63.2\%$ of its asymptotic value, or 5.69 V in this case. From the graph, $\tau \simeq 1.5$ ms. (c) The time constant is RC, so $C = 1.5\ \text{ms}/4700\ \Omega = 0.319\ \mu\text{F}$.

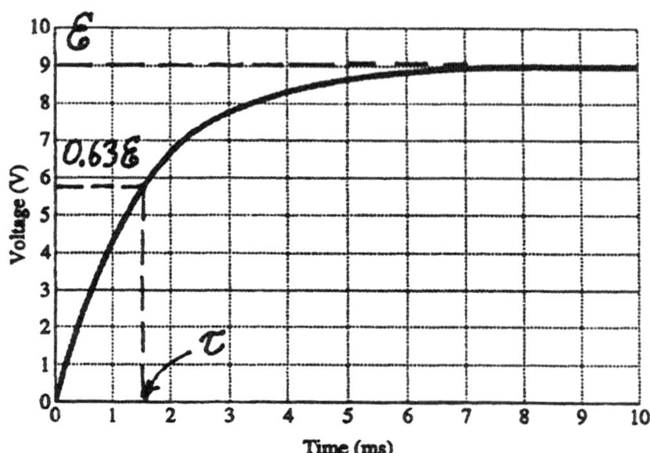

FIGURE 28-62 Problem 49 Solution.

Problem

51. A 1.0-μF capacitor is charged to 10.0 V. It is then connected across a 500-kΩ resistor. How long does it take (a) for the capacitor voltage to reach 5.0 V and (b) for the energy stored in the capacitor to decrease to half its initial value?

Solution

A capacitor discharging through a resistor is described by exponential decay, with time constant RC (see Equation 28-8), and, of course, $U_C(t) = \frac{1}{2}CV(t)^2 = \frac{1}{2}CV_0^2e^{-2t/RC} = U_C(0)e^{-2t/RC}$ is the energy stored (see Equation 26-8b). (a) $V(t)/V(0) = 1/2$ implies $t = RC\ln 2 = (500\text{ k}\Omega)(1\ \mu\text{F})(0.693) = 347$ ms. (b) $U_c(t)/U_c(0) = 1/2$ implies $t = \frac{1}{2}RC\ln 2 = 173$ ms.

Problem

53. A capacitor is charged until it holds 5.0 J of energy. It is then connected across a 10-kΩ resistor. In 8.6 ms, the resistor dissipates 2.0 J. What is the capacitance?

Solution

Equation 28-8 gives a voltage $V = V_0e^{-t/RC}$ for a capacitor discharging through a resistor. If 2 J is dissipated in time t, the energy stored in the capacitor drops from $U_0 = 5$ J to $U = 3$ J (assuming there are no losses due to radiation, etc.). Since $U = \frac{1}{2}CV^2$, $U_0/U = (V_0/V)^2 = e^{2t/RC}$ and we may solve for C: $C = 2t/R\ln(U_0/U) = 2(8.6\text{ ms})/(10\text{ k}\Omega)\times \ln(5/3) = 3.37\ \mu$F.

Problem

55. For the circuit of Example 28-9, take $\mathcal{E} = 100$ V, $R_1 = 4.0$ kΩ, and $R_2 = 6.0$ kΩ, and assume the capacitor is initially uncharged. What are the currents in both resistors and the voltage across the capacitor (a) just after the switch is closed and (b) a long time after the switch is closed? Long after the switch is closed it is again opened. What are I_1, I_2, and V_C (c) just after this switch opening and (d) a long time later?

Solution

In addition to the explanation in Example 28-9, we note that when the switch is in the closed position, Kirchhoff's voltage law applied to the loop containing both resistors yields $\mathcal{E} = I_1R_1 + I_2R_2$, and to the loop containing just R_2 and C, $V_C = I_2R_2$. (a) If the switch is closed at $t = 0$, Example 28-9 shows that $V_C(0) = 0$, $I_2(0) = 0$, and $I_1(0) = \mathcal{E}/R_1 = 100\text{ V}/4\text{ k}\Omega = 25$ mA. (b) After a long time, $t = \infty$, Example 28-9 also shows that $I_1(\infty) = I_2(\infty) = \mathcal{E}/(R_1 + R_2) = 100\text{ V}/10\text{ k}\Omega = 10$ mA, and $V_C(\infty) = I_2(\infty)R_2 = (10\text{ mA})(6\text{ k}\Omega) = 60$ V. (c) Under the conditions stated, the fully charged capacitor ($V_C = 60$ V) simply discharges through R_2. (R_1 is in an open-circuit branch, so $I_1 = 0$ for the entire discharging process.) The initial discharging current is $I_2 = V_C/R_2 = 60\text{ V}/6\text{ k}\Omega = 10$ mA. (d) I_2 and V_C decay exponentially to zero.

Problem

57. In the circuit for Fig. 28-65 the switch is initially open and the capacitor is uncharged. Find expressions for the current I supplied by the battery (a) just after the switch is closed and (b) a long time after the switch is closed.

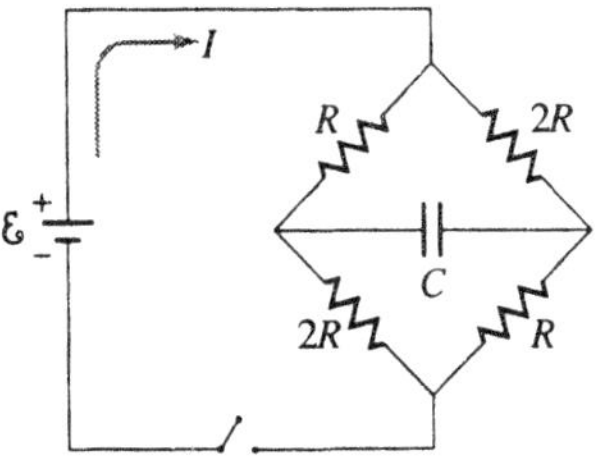

FIGURE 28-65 Problem 57.

Solution

(a) Just after the switch is closed, the uncharged capacitor acts instantaneously like a short circuit and the resistors act like two parallel pairs in series. The effective resistance of the combination is $2\times(R)(2R)/(R + 2R) = 4R/3$, and the current supplied by the battery is $I(0) = 3\mathcal{E}/4R$. (b) A long time after the switch is closed, the capacitor is fully charged and acts like an open circuit. Then the resistors act like two series pairs in parallel, with an effective resistance of $(\frac{1}{2})(R + 2R) = 3R/2$. The battery current is $I(\infty) = 2\mathcal{E}/3R$.

Paired Problems

Problem

59. A 3.3-kΩ resistor and a 4.7-kΩ resistor are connected in parallel, and the pair is in series with a 1.5-kΩ resistor. What is the resistance of the combination?

Solution

Equation 28-3c for the resistance of the parallel pair, combined with Equation 28-1 for the resistance of this in series with the third resistor gives an effective resistance of $R_{\text{eff}} = 1.5\text{ k}\Omega + (3.3)(4.7)\text{ k}\Omega/(3.3 + 4.7) = 3.44$ kΩ.

Problem

61. A battery's voltage is measured with a voltmeter whose resistance is 1000 Ω; the result is 4.36 V. When the measurement is repeated with a 1500-Ω meter the result is 4.41 V. What are (a) the battery voltage and (b) its internal resistance?

Solution

The internal resistance of the battery (R_i) and the resistance of the voltmeter (R_m) are in series with the battery's emf, so the current is $I = \mathcal{E}/(R_i + R_m)$. The potential drop across the meter (its reading) is $V_m = IR_m = \mathcal{E}R_m/(R_i + R_m)$. From the given data, $4.36\ \text{V} = \mathcal{E}(1\ \text{k}\Omega)/(R_i + 1\ \text{k}\Omega)$ and $4.41\ \text{V} = \mathcal{E}(1.5\ \text{k}\Omega)/(R_i + 1.5\ \text{k}\Omega)$, which can be solved simultaneously for $\mathcal{E}$ and R_i. One obtains $R_i + 1\ \text{k}\Omega = \mathcal{E}(1\ \text{k}\Omega/4.36\ \text{V})$ and $R_i + 1.5\ \text{k}\Omega = \mathcal{E}(1.5\ \text{k}\Omega/4.41\ \text{V})$, or

$$\mathcal{E} = (1.5\ \text{k}\Omega - 1\ \text{k}\Omega)\left(\frac{1.5\ \text{k}\Omega}{4.41\ \text{V}} - \frac{1\ \text{k}\Omega}{4.36\ \text{V}}\right)^{-1} = 4.51\ \text{V}$$

and $R_i = (4.51\ \text{V})(1\ \text{k}\Omega/4.36\ \text{V}) - 1\ \text{k}\Omega = 35.2\ \Omega$.

Problem

63. In Fig. 28-67, take $\mathcal{E}_1 = 12$ V, $\mathcal{E}_2 = 6.0$ V, $\mathcal{E}_3 = 3.0$ V, $R_1 = 1.0\ \Omega$, $R_2 = 2.0\ \Omega$, and $R_3 = 4.0\ \Omega$. Find the current in R_2 and give its direction.

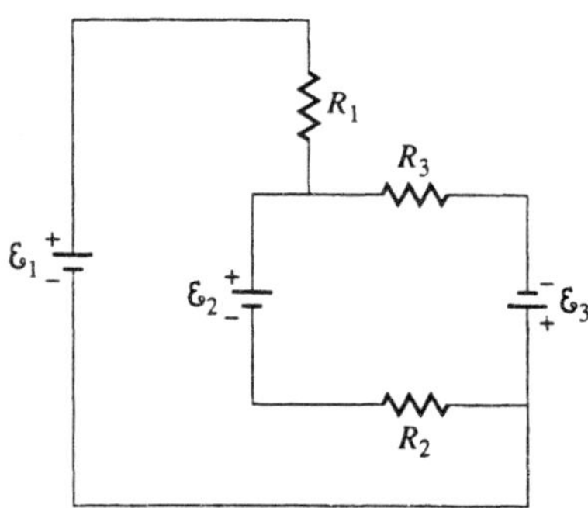

FIGURE 28-67 Problem 63.

Solution

An obvious reconfiguration of the circuit in Fig. 28-67 results in a circuit like that in Fig. 28-57, with R_1 replacing $(R_1 + R_2)$, R_2 for R_3, R_3 for R_4, and $-\mathcal{E}_3$ for $\mathcal{E}_3$. Thus, the solution to Problem 35, properly altered, gives

$$I_b = \frac{(\mathcal{E}_2 - \mathcal{E}_1)R_3 + (\mathcal{E}_2 + \mathcal{E}_3)R_1}{R_1R_2 + R_1R_3 + R_2R_3}$$

$$= \frac{(6 - 12)(4) + (6 + 3)(1)}{(1)(2 + 4) + (2)(4)}\ \text{A} = -1.07\ \text{A}.$$

(A negative current is opposite to the direction of the emf $\mathcal{E}_2$.)

Problem

65. In Fig. 28-68 what are the meter readings when (a) an ideal voltmeter or (b) an ideal ammeter is connected between points A and B?

Solution

(a) An ideal voltmeter has $R_m = \infty$ (AB open circuited), so V_{AB} is just the voltage across the 5.6 kΩ resistor. This is part of a voltage divider (in series with the 4.7 kΩ resistor), so Equation 28-2 gives $V_{AB} = (24\ \text{V})(5.6)/(5.6 + 4.7) = 13.0$ V. (b) An ideal ammeter has $R_m = 0$ (AB short circuited), so I_{AB} is just the current through the 3.3 kΩ resistor. This is part of parallel combination, $R_{||} = (3.3\ \text{k}\Omega)(5.6)\div(3.3 + 5.6) = 2.08\ \text{k}\Omega$, which, in series with the 4.7 kΩ resistor, draws a total current of $I_{\text{tot}} = 24\ \text{V}\div(2.08 + 4.7)\ \text{k}\Omega = 3.54$ mA. Now, two resistors in parallel form a current divider, each one taking a fraction of the total current, given by $I_1 = (R_{||}/R_1)I_{\text{tot}}$, and $I_2 = (R_{||}/R_2)I_{\text{tot}}$, respectively. (This follows directly from Ohm's law: $V_{||} = I_{\text{tot}}R_{||} = I_1R_1 = I_2R_2$.) Therefore, in the case of the ideal ammeter, $I_{AB} = (R_{||}/3.3\ \text{k}\Omega)I_{\text{tot}} = (2.08/3.3)\times(3.54\ \text{mA}) = 2.23$ mA.

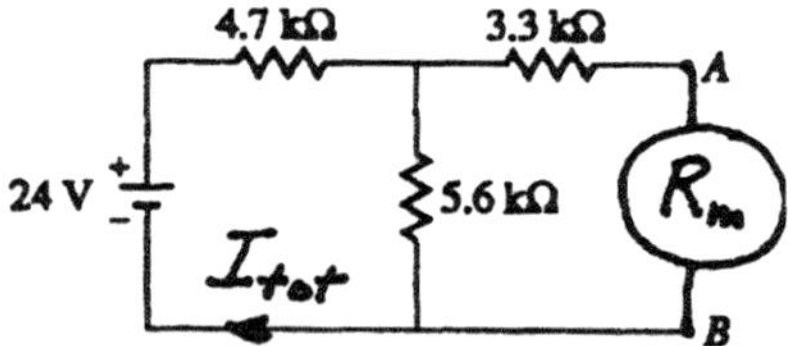

FIGURE 28-68 Problem 65 Solution.

Problem

67. An initially uncharged capacitor in an RC circuit reaches 75% of its full charge in 22.0 ms. What is the time constant?

Solution

From Equation 28-6, $V_C/\mathcal{E} = 75\% = 1 - e^{-t/\tau}$, which implies $e^{t/\tau} = 4$, or $\tau = t/\ln 4 = 22\ \text{ms}/\ln 4 = 15.9$ ms.

Supplementary Problems

Problem

69. Suppose the currents into and out of a circuit node differed by 1 μA. If the node consists of a small metal sphere with diameter 1 mm, how long would it take for the electric field around the node to reach the breakdown field in air (3 MV/m)?

Solution

The charge on the node (whether positive or negative) accumulates at a rate of 1 μA = 1 μC per second, so $|q(t)| = (1\ \mu\text{A})t$ (where we assume that $q(0) = 0$). If the node is treated approximately as an isolated sphere, the electric field strength at its surface, $k\,|q|\,/r^2 = k(1\ \mu\text{A})t/r^2$, equals the breakdown field for air, when

$$t = (3\ \text{MV/m})(0.5\ \text{mm})^2/(9\times10^9\ \ \text{m/F})(1\ \mu\text{A})$$
$$= 83.3\ \mu\text{s}.$$

Problem

71. In Fig. 28-70, what is the current in the 4-Ω resistor when each of the following circuit elements is connected between points A and B. (a) an ideal ammeter; (b) an ideal voltmeter; (c) another 4.0-Ω resistor; (d) an uncharged capacitor, right after it's connected; (e) long after the capacitor of part (d) is connected; (f) an ideal 12-V battery, with its positive terminal at A; (g) a capacitor initially charged to 12 V, right after it's connected with its positive plate at A; (h) long after the capacitor in part (g) is connected?

Solution

(a) An ideal ammeter connected across AB short-circuits the 4 Ω resistor, so $I_4 = 0$. (This is a dangerous connection unless the ammeter can handle 3 A.) (b) An ideal voltmeter across AB maintains the open circuit, so $I_4 = 6\ \text{V}/(2+4)\ \Omega = 1$ A. (c) Another 4 Ω resistor in parallel across AB divides the total current equally; $I_{\text{tot}} = 6\ \text{V}/[2+4\times4/(4+4)]\ \Omega =$

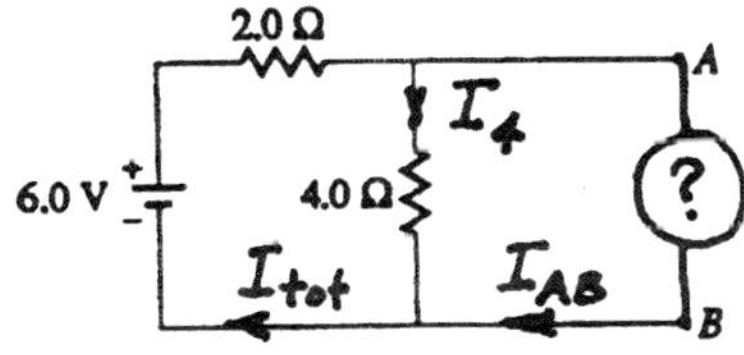

FIGURE 28-70 Problem 71 Solution.

1.5 A, and $I_4 = 0.75$ A. (d) The instantaneous voltage across an uncharged capacitor is zero, so AB is momentarily short-circuited and $I_4 = 0$, as in (a). (e) When the capacitor is fully charged ($dq/dt = 0$) it behaves like an open circuit, so $I_4 = 1$ A, as in (b). (f) Kirchhoff's voltage law applied to the loop containing just AB (with the ideal 12 V emf) and the 4 Ω resistor, gives 12 V $= I_4(4\ \Omega)$, or $I_4 = 3$ A. (g) Instantaneously, a capacitor charged to 12 V appears like the battery in (f) when first connected, so $I_4 = 3$ A. (h) Charge leaves the capacitor until, after a long time, $dq/dt = 0 = I_C$ and it appears like an open circuit. Then $I_4 = 1$ A, as in (b) and (e).

Problem

73. A parallel-plate capacitor is insulated with a material of dielectric constant κ and resistivity ρ. Since the resistivity is finite, the capacitor "leaks" charge and can be modeled as an ideal capacitor in parallel with a resistor. (a) Show that the time constant of the capacitor is independent of its dimensions (provided the spacing is small enough that the usual parallel-plate approximation applies) and is given by $\varepsilon_0\kappa\rho$. (b) If the insulating material is polystyrene ($\kappa = 2.6$, $\rho = 10^{16}$ Ω·m), how long will it take for the stored energy in the capacitor to decrease by a factor of 2?

Solution

(a) In the parallel plate approximation, the electric field between the plates is constant, $E = V/d$, the leakage current density is uniform, $J = E/\rho$, and the capacitance is $C = \kappa\varepsilon_0 A/d$. These relations imply that the resistance of the dielectric slab between the plates is $R = V/I = V/JA = V/(EA/\rho) = \rho d/A$, and the time constant for discharging through the dielectric is $\tau = RC = (\rho d/A)(\kappa\varepsilon_0 A/d) = \kappa\varepsilon_0\rho$. (b) The stored energy in the capacitor, $U_C = \frac{1}{2}CV_C^2$, decays with half the voltage time constant, i.e., $U_C = \frac{1}{2}C(V_0e^{-t/RC})^2 = \frac{1}{2}CV_0^2e^{-2t/RC} = U_0e^{-t/(RC/2)}$. To decay by 50% takes time $t = (RC/2)\ln\ 2 = (\kappa\varepsilon_0\rho/2)\ln 2$. With the values given for polystyrene, $t = (2.6)(8.85\ p\text{F/m})(10^{16}\ \Omega\cdot\text{m})\frac{1}{2}\ln 2 = 7.97\times10^4\ \text{s} = 22.2$ h.

Problem

75. Write the loop and node laws for the circuit of Fig. 28-71, and show that the time constant for this circuit is $R_1R_2C/(R_1+R_2)$.

Solution

Consider the loops and node added to Fig. 28-71. Kirchhoff's laws are $\mathcal{E} = I_1R_1 + I_2R_2$, $V_C = I_2R_2$, and $I_C = I_1 - I_2$. Since $V_C = q/C$ and $I_C = dq/dt$, the equations can be combined to yield

$$\begin{aligned}\mathcal{E} - I_1R_1 - I_2R_2 &= \mathcal{E} - (I_C + I_2)R_1 - I_2R_2 \\ &= \mathcal{E} - I_CR_1 - \left(\frac{V_C}{R_2}\right)(R_1 + R_2) \\ &= \mathcal{E} - I_CR_1 - \frac{q}{CR_2/(R_1 + R_2)} = 0.\end{aligned}$$

This is exactly in the same form as the first equation, solved in the text, in the section "The RC Circuit: Charging" (with $I \to I_C$, $R \to R_1$ and $C \to CR_2 \div (R_1 + R_2)$), so the time constant for the circuit is $\tau = CR_1R_2/(R_1 + R_2)$ (the ratio of the coefficients of I_C and q).

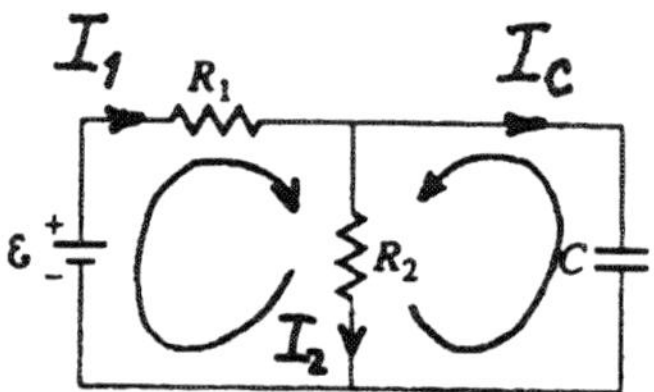

FIGURE 28-71 Problem 75 Solution.

CHAPTER 29 THE MAGNETIC FIELD

ActivPhysics can help with these problems: Activities 13.4, 13.6, 13.7, 13.8

Section 29-2: The Magnetic Force and Moving Charge

Problem

1. (a) What is the minimum magnetic field needed to exert a 5.4×10^{-15}-N force on an electron moving at 2.1×10^7 m/s? (b) What magnetic field strength would be required if the field were at 45° to the electron's velocity?

Solution

(a) From Equation 29-1b, $B = F/ev\sin\theta$, which is a minimum when $\sin\theta = 1$ (the magnetic field perpendicular to the velocity). Thus, $B_{\text{min}} = (5.4\times10^{-15}\text{ N})/(1.6\times10^{-19}\text{ C})(2.1\times10^7\text{ m/s}) = 1.61\times10^{-3}\text{ T} = 16.1$ G. (b) For $\theta = 45°$, $B = B_{\text{min}}\div\sin 45° = \sqrt{2}\,B_{\text{min}} = 22.7$ G.

Problem

3. What is the magnitude of the magnetic force on a proton moving at 2.5×10^5 m/s (a) at right angles; (b) at 30°; (c) parallel to a magnetic field of 0.50 T?

Solution

From Equation 29-1b, $F = evB\sin\theta$, so (a) when $\theta = 90°$, $F = (1.6\times10^{-19}\text{ C})(2.5\times10^5\text{ m/s})(0.5\text{ T}) = 2.0\times10^{-14}$ N, (b) $F = (2.0\times10^{-14}\text{ N})\sin 30° = 1.0\times10^{-14}$ N, and (c) $F = evB\sin 0° = 0$.

Problem

5. A particle carrying a 50-μC charge moves with velocity $\mathbf{v} = 5.0\hat{\imath} + 3.2\hat{\mathbf{k}}$ m/s through a uniform magnetic field $\mathbf{B} = 9.4\hat{\imath} + 6.7\hat{\jmath}$ T. (a) What is the force on the particle? (b) Form the dot products $\mathbf{F}\cdot\mathbf{v}$ and $\mathbf{F}\cdot\mathbf{B}$ to show explicitly that the force is perpendicular to both $\mathbf{v}$ and $\mathbf{B}$.

Solution

(a) From Equation 27-2, $\mathbf{F} = q\mathbf{v}\times\mathbf{B} = (50\ \mu\text{C})\times(5\hat{\imath} + 3.2\hat{\mathbf{k}}\text{ m/s})(9.4\hat{\imath} + 6.7\hat{\jmath}\text{ T}) = (50\times10^{-6}\text{ N})\times(5\times6.7\hat{\mathbf{k}} + 3.2\times9.4\hat{\jmath} - 3.2\times6.7\hat{\imath}) = (-1.072\hat{\imath} + 1.504\hat{\jmath} + 1.675\hat{\mathbf{k}})\times10^{-3}$ N. (The magnitude and direction can be found from the components, if desired.) (b) The dot products $\mathbf{F}\cdot\mathbf{v}$ and $\mathbf{F}\cdot\mathbf{B}$ are, respectively, proportional to $(-1.072)(5) + (1.675)\times(3.2) = 0$, and $(-1.072)(9.4) + (1.504)(6.7) = 0$, since the cross product of two vectors is perpendicular to each factor. (We did not round off the components of $\mathbf{F}$, so that the vanishing of the dot products could be exactly confirmed.)

Problem

7. A proton moving with velocity $\mathbf{v}_1 = 3.6\times10^4\ \hat{\jmath}$m/s experiences a magnetic force of $7.4\times10^{-16}\ \hat{\imath}$N. A second proton moving on the x axis experiences a magnetic force of $2.8\times10^{-16}\ \hat{\jmath}$N. Find the magnitude and direction of the magnetic field, and the velocity of the second proton.

Solution

The magnetic force on the first proton is $(7.4\times10^{-16}\text{ N})\hat{\imath} = e(v_1\hat{\jmath})(B_x\hat{\imath} + B_y\hat{\jmath} + B_z\hat{\mathbf{k}}) = ev_1(-B_x\hat{\mathbf{k}} + B_z\hat{\imath})$, so $B_x = 0$ and $7.4\times10^{-16}\text{ N} = ev_1B_z$. The force on the second proton is $(2.8\times10^{-16}\text{ N})\hat{\jmath} = e(v_2\hat{\imath})\times(B_y\hat{\jmath} + B_z\hat{\mathbf{k}}) = ev_2(B_y\hat{\mathbf{k}} - B_z\hat{\jmath})$, so $B_y = 0$ and $2.8\times10^{-16}\text{N} = -ev_2B_z$. Therefore, $\mathbf{B} = B_z\hat{\mathbf{k}} = \hat{\mathbf{k}}(7.4\times10^{-16}\text{ N})\div(1.6\times10^{-19}\text{ C})(3.6\times10^4\text{ m/s}) = (0.128\text{ T})\hat{\mathbf{k}}$, and $\mathbf{v}_2 = v_2\hat{\imath} = \hat{\imath}(-2.8/7.4)(3.6\times10^4\text{ m/s}) = (-1.36\times10^4\text{ m/s})\hat{\imath}$.

Problem

9. An alpha particle (2 protons, 2 neutrons) is moving with velocity $\mathbf{v} = 150\hat{\imath} + 320\hat{\jmath} - 190\hat{\mathbf{k}}$ km/s in a magnetic field is $\mathbf{B} = 0.66\hat{\imath} - 0.41\hat{\jmath}$ T. Find the magnitude of the force on the particle.

Solution

The magnetic force on the alpha particle is (Equation 29-1a): $\mathbf{F}_B = 2e\mathbf{v}\times\mathbf{B} = 2(1.6\times10^{-19}\text{ C})\times(150\hat{\imath} + 320\hat{\jmath} - 190\hat{\mathbf{k}})(10^3\text{ m/s})\times(0.66\hat{\imath} - 0.41\hat{\jmath})\text{ T} = (3.2\times10^{-16}\text{ N})(-150\times0.41\hat{\mathbf{k}} - 320\times0.66\hat{\mathbf{k}} - 190\times0.66\hat{\jmath} - 190\times0.41\hat{\imath}) = -(24.9\hat{\imath} + 40.1\hat{\jmath} + 87.3\hat{\mathbf{k}})$ fN.

Problem

11. A 1.4-μC charge moving at 185 m/s experiences a magnetic force $\mathbf{F}_B = 2.5\hat{\imath} + 7.0\hat{\jmath}\ \mu$N in a magnetic field $\mathbf{B} = 4.2\hat{\imath} - 15\hat{\jmath}$ mT. What is the angle

between the particle's velocity and the magnetic field?

Solution

Equation 29-1b gives $\sin\theta = F/qvB = \sqrt{(2.5^2)+(7.0)^2}\ \mu\text{N} \div (1.4\ \mu\text{C})(185\ \text{m/s})\times \sqrt{(42^2)+(-15)^2}\ \text{mT} = 0.644$. Then $\theta = 40.1°$ or $140°$ (both are possible since $\sin\theta = \sin(180° - \theta)$).

Problem

13. A region contains an electric field $\mathbf{E} = 7.4\hat{\imath} + 2.8\hat{\jmath}$ kN/C and a magnetic field $\mathbf{B} = 15\hat{\jmath} + 36\hat{\mathbf{k}}$ mT. Find the electromagnetic force on (a) a stationary proton, (b) an electron moving with velocity $\mathbf{v} = 6.1\hat{\imath}$ Mm/s.

Solution

The force on a moving charge is given by Equation 29-2 (called the Lorentz force) $\mathbf{F} = q(\mathbf{E} + \mathbf{v}\times\mathbf{B})$. (a) For a stationary proton, $q = e$ and $\mathbf{v} = 0$, so $\mathbf{F} = e\mathbf{E} = (1.6\times10^{-19}\ \text{C})(7.4\hat{\imath} + 2.8\hat{\jmath})\ \text{kN/C} = (1.18\hat{\imath} + 0.448\hat{\jmath})$ fN. (b) For the electron, $q = -e$ and $\mathbf{v} = 6.1\hat{\imath}$ Mm/s, so the electric force is the negative of the force in part (a) and the magnetic force is $-e\mathbf{v}\times\mathbf{B} = (-1.6\times10^{-19}\ \text{C})(6.1\hat{\imath}\ \text{Mm/s})\times(15\hat{\jmath} + 36\hat{\mathbf{k}})\ \text{mT} = (-14.6\hat{\mathbf{k}} + 35.1\hat{\jmath})$ fN. The total Lorentz force is the sum of these, or $(-1.18\hat{\imath} + 34.7\hat{\jmath} - 14.6\hat{\mathbf{k}})$ fN.

Section 29-3: The Motion of Charged Particles in Magnetic Fields

Problem

15. What is the radius of the circular path described by a proton moving at 15 km/s in a plane perpendicular to a 400-G magnetic field?

Solution

From Equation 29-3, the radius of the orbit is $r = mv/eB = (1.67\times10^{-27}\ \text{kg})(15\ \text{km/s})/(1.6\times10^{-19}\ \text{C})\times(4\times10^{-2}\ \text{T}) = 3.91$ mm. (SI units and data for the proton are summarized in the appendices and inside front cover.)

Problem

17. Radio astronomers detect electromagnetic radiation at a frequency of 42 MHz from an interstellar gas cloud. If this radiation is caused by electrons spiraling in a magnetic field, what is the field strength in the gas cloud?

Solution

If this is electromagnetic radiation at the electron's cyclotron frequency, Equation 29-5 implies a field strength of $B = 2\pi f(m/e) = 2\pi(42\ \text{MHz})\times(9.11\times10^{-31}\ \text{kg}/1.6\times10^{-19}\ \text{C}) = 1.50\times10^{-3}\ \text{T} = 15.0$ G.

Problem

19. Electrons and protons with the same kinetic energy are moving at right angles to a uniform magnetic field. How do their orbital radii compare?

Solution

It is convenient to anticipate the result of Problem 22 for the orbital radius of a non-relativistic charged particle in a plane perpendicular to a uniform magnetic field. From Equation 29-3, $r = mv/qB$. For a non-relativistic particle, $K = \frac{1}{2}mv^2$, or $v = \sqrt{2K/m}$, therefore $r = \sqrt{2Km}/qB$. (Note: All quantities can, of course, be expressed in standard SI units, but in many applications, atomic units are more convenient. The conversion factor for electron volts to joules is just the numerical magnitude of the electronic charge, so if K is expressed in MeV, m in MeV/c^2, q in multiples of e, and B in teslas, we obtain

$$r = \frac{\sqrt{2K(e\times10^6)m(e\times10^6/c^2)}}{(qe)B}$$

$$= \frac{10^6\sqrt{2Km}}{(3\times10^8)qB} = \frac{\sqrt{2Km}}{300qB}.\Bigg)$$

From this expression, it follows that protons and electrons with the same kinetic energy have radii in the ratio $r_p/r_e = \sqrt{m_p/m_e} = \sqrt{1836} \approx 43$, in the same magnetic field. Heavier particles are more difficult to bend.

Problem

21. Microwaves in a microwave oven are produced by electrons circling in a magnetic field at a frequency of 2.4 GHz. (a) What is the magnetic field strength? (b) The electrons' motion takes place inside a special tube called a magnetron. If the magnetron can accommodate electron orbits with a maximum diameter of 2.5 mm, what is the maximum electron energy?

Solution

(a) A cyclotron frequency of 2.4 GHz for electrons implies a magnetic field strength of $B = 2\pi f(m/e) = 2\pi(2.4\ \text{GHz})(9.11\times10^{-31}\ \text{kg}/1.6\times10^{-19}\ \text{C}) = 85.9$ mT (see Equation 29-5). (b) The kinetic energy of an electron, with the maximum orbital radius allowed for this magnetron tube (half the diameter), in the field found in part (a), is $K = (reB)^2/2m = (1.25\ \text{mm}\ \times$

1.6×10^{-19} C $\times$ 85.9 mT$)^2/(2\times9.11\times10^{-31}$ kg$)$ = $(1.62\times10^{-16}$ J$)/(1.6\times10^{-19}$ J/eV$)$ = 1.01 keV (see Problem 22). The same calculation in atomic units, explained in the solution to Problem 19, is $(1.25\times10^{-3}\times300\times85.9\times10^{-3})^2(2\times0.511)^{-1}$ MeV. The electron's kinetic energy could also be expressed in terms of the cyclotron frequency directly, $K = (2\pi frm)^2/2m = 2m(\pi rf)^2$, with the same result.

Problem

23. Two protons, moving in a plane perpendicular to a uniform magnetic field of 500 G, undergo an elastic head-on collision. How much time elapses before they collide again? *Hint:* Draw a picture.

Solution

In an elastic head on collision between particles of equal mass, the particles exchange velocities ($v_{1f} = v_{2i}$ and $v_{2f} = v_{1i}$, see Section 11-4). Moving in a plane perpendicular to **B**, each proton describes a different circle of radius $r = mv/eB$, with period (which is independent of r and v) of $T = 2\pi m/eB = 2\pi(1.67\times10^{-27}\text{ kg})/(1.6\times10^{-19}\text{ C})(5\times10^{-2}\text{ T}) =$ 1.31 μs. After one period, each proton would be back at the site of the collision, and could collide again.

Problem

25. A cyclotron is designed to accelerate deuterium nuclei. (Deuterium has one proton and one neutron in its nucleus.) (a) If the cyclotron uses a 2.0-T magnetic field, at what frequency should the dee voltage be alternated? (b) If the vacuum chamber has a diameter of 0.90 m, what is the maximum kinetic energy of the deuterons? (c) If the magnitude of the potential difference between the dees is 1500 V, how many orbits do the deuterons complete before achieving the energy of part (b)?

Solution

(a) The frequency of the accelerating voltage is the cyclotron frequency for deuterons (Equation 29-5), $f = eB/2\pi m \simeq (1.6\times10^{-19}\text{ C})(2\text{ T})/2\pi(2\times1.67\times10^{-27}\text{ kg}) = 15.2$ MHz. (b) We can use the result of Problem 19 (expressed in atomic units), with the maximum orbital radius equal to the radius of the dees. Thus, $K_{\text{max}} = (300qBr_{\text{max}})^2/2m \simeq (300\times1\times2\times0.45)^2/2(2\times938) = 19.4$ MeV. (c) If the deuterons start with essentially zero kinetic energy, and gain 1500 eV each half-orbit, they will make 19.4 MeV/2(1500 eV) = 6.48×10^3 orbits. (Of course, the same results follow in standard SI units.)

Problem

27. Figure 29-38 shows a simple mass spectrometer, designed to analyze and separate atomic and molecular ions with different charge-to-mass ratios. In the design shown, ions are accelerated through a potential difference V, after which they enter a region containing a uniform magnetic field. They describe semicircular paths in the magnetic field, and land on a detector a lateral distance x from where they entered the field region, as shown. Show that x is given by

$$x = \frac{2}{B}\sqrt{\frac{2V}{(q/m)}},$$

where B is the magnetic field strength, V the accelerating potential, and q/m the charge-to-mass ratio of the ion. By counting the number of ions accumulated at different positions x, one can determine the relative abundances of different atomic or molecular species in a sample.

Solution

The positive ions enter the field region with speed (determined from the work-energy theorem) of $\frac{1}{2}mv^2 = qV$, or $v = \sqrt{2V(q/m)}$. They are bent into a semicircle with diameter $x = 2r = 2mv/qB = 2(m/qB)\sqrt{2V(q/m)} = 2\sqrt{2(m/q)V}/B$, as shown in Fig. 29-38 (see Equation 29-3).

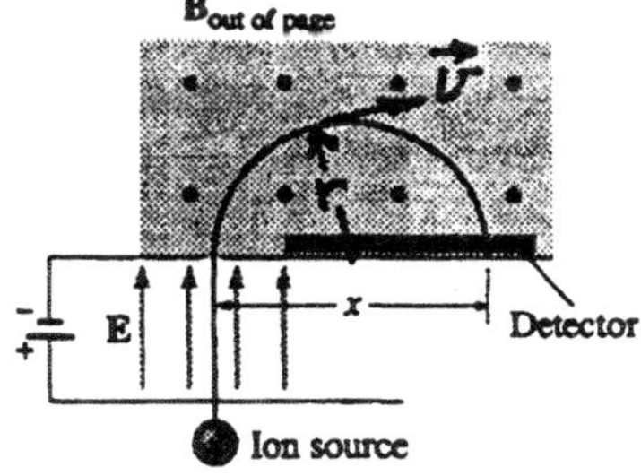

FIGURE 29-38 Problem 27 Solution.

Problem

29. A mass spectrometer is used to separate the fissionable uranium isotope U-235 from the much more abundant isotope U-238. To within what percentage must the magnetic field be held constant if there is to be no overlap of these two isotopes? Both isotopes appear as constituents of uranium hexafluoride gas (UF_6), and the gas molecules are all singly ionized.

Solution

The separation of different uranium isotopes in UF_6 molecules can be found by differentiation of the result of Problem 27. Keeping the spectrometer parameters fixed, $x \sim m^{1/2}$, $dx \sim \frac{1}{2}m^{-1/2}dm$, and $dx/x = \frac{1}{2}(dm/m)$. The molecular masses of the two species are approximately 235 or 238 plus 6×19 which equals 349 or 352, respectively, so $\frac{1}{2}(dm/m) \approx 3/2\times350 = 0.43\%$. For a particular ion, $x \sim B^{-1}$ and $dx \sim B^{-2}dB$, therefore $dx/x = -dB/B$. Thus, variations in B should be less than 0.43% to separate these isotopes.

Problem

31. An electron moving at 3.8×10^6 m/s enters a region containing a uniform magnetic field $\mathbf{B} = 18\hat{\mathbf{k}}$ mT. The electron is moving at 70° to the field direction, as shown in Fig. 29-39. Find the radius r and pitch p of its spiral path, as indicated in the figure.

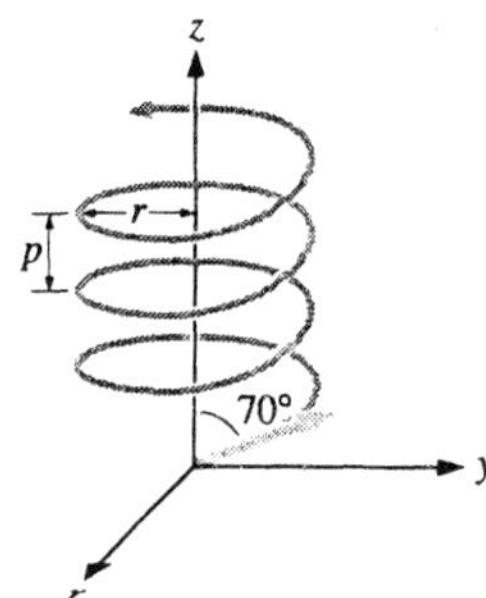

FIGURE 29-39 Problem 31.

Solution

The parallel and perpendicular components of the electron's velocity (relative to the field direction, or z axis) are $v_{\|} = v\cos 70^\circ = (3.8\times10^6 \text{ m/s})\cos 70^\circ = 1.30$ Mm/s, and $v_\perp = v\sin 70^\circ = 3.57$ Mm/s. Then $r = mv_\perp/eB = 1.13$ mm and the pitch $p = v_{\|}T = v_{\|}(2\pi m/eB) = 2.58$ mm.

Section 29-4: The Magnetic Force on a Current

Problem

33. What is the magnitude of the force on a 50-cm-long wire carrying 15 A at right angles to a 500-G magnetic field?

Solution

The force on a straight current-carrying wire in a uniform magnetic field is (Equation 29-6) $\mathbf{F} = I\boldsymbol{\ell}\times\mathbf{B}$. Thus, $F = I\ell B\sin\theta = (15 \text{ A})(0.5 \text{ m})(0.05 \text{ T})\sin 90^\circ = 0.375$ N. (The direction is given by the right-hand rule.)

Problem

35. A wire carrying 15 A makes a 25° angle with a uniform magnetic field. The magnetic force per unit length of wire is 0.31 N/m. (a) What is the magnetic field strength? (b) What is the maximum force per unit length that could be achieved by reorienting the wire in this field?

Solution

Equation 29-6 gives the magnetic force on a straight current carrying wire in a uniform magnetic field, $\mathbf{F} = I\boldsymbol{\ell}\times\mathbf{B}$. (a) From the magnitude of $\mathbf{F}$ and the given data, we find $B = F/I\ell\sin\theta = (0.31 \text{ N/m})\div(15 \text{ A})\sin 25^\circ = 48.9$ mT. (b) By placing the wire perpendicular to the field ($\sin\theta = 1$) a maximum force per unit length of $IB = (15 \text{ A})(48.9 \text{ mT}) = 0.734$ N/m could be attained.

Problem

37. In a high-magnetic-field experiment, a conducting bar carrying 7.5 kA passes through a 30-cm-long region containing a 22-T magnetic field. If the bar makes a 60° angle with the field direction, what force is necessary to hold it in place?

Solution

The magnitude of the force necessary to balance the magnetic force on the bar (Equation 29-6) is $F = |I\boldsymbol{\ell}\times\mathbf{B}| = I\ell B\sin\theta = (7.5 \text{ kA})(0.3 \text{ m})(22 \text{ T})\times\sin 60^\circ = 42.9$ kN (nearly 5 tons). The direction of this force is perpendicular to the plane of $\boldsymbol{\ell}\times\mathbf{B}$ in the opposite sense as the magnetic force.

Problem

39. A piece of wire with mass per unit length 75 g/m runs horizontally at right angles to a horizontal magnetic field. A 6.2-A current in the wire results in its being suspended against gravity. What is the magnetic field strength?

Solution

A magnetic force equal in magnitude to the weight of the wire requires that $I\ell B = mg$ (since the wire is perpendicular to the field), or $B = (m/\ell)(g/I) = (75 \text{ g/m})(9.8 \text{ m/s}^2)/(6.2 \text{ A}) = 0.119$ T.

Problem

41. A wire carrying 1.5 A passes through a region containing a 48-mT magnetic field. The wire is perpendicular to the field and makes a quarter-circle turn of radius 21 cm as it passes through the field region, as shown in Fig. 29-43.

Find the magnitude and direction of the magnetic force on this section of wire.

Solution

Take the x-y axes as shown on Fig. 29-43, with the z axis out of the page and the origin at the center of the quarter-circle arc. With θ measured clockwise from the y axis, $d\boldsymbol{\ell} = R\,d\theta(\hat{\imath}\cos\theta - \hat{\jmath}\sin\theta)$ as shown, and $\mathbf{B} = B(-\hat{\mathbf{k}})$. The magnetic force on the arc of wire is found from Equation 29-8.

$$\mathbf{F} = I\int_{arc} d\boldsymbol{\ell}\times\mathbf{B}$$
$$= I\int_0^{90^\circ} R\,d\theta(\hat{\imath}\cos\theta - \hat{\jmath}\sin\theta)\times B(-\hat{\mathbf{k}})$$
$$= IRB\int_0^{90^\circ} (\hat{\jmath}\cos\theta + \hat{\imath}\sin\theta)\,d\theta$$
$$= IRB|-\hat{\imath}\cos\theta + \hat{\jmath}\sin\theta|_0^{90^\circ} = IRB(\hat{\imath}+\hat{\jmath}).$$

This has magnitude $\sqrt{2}\,IRB = \sqrt{2}(1.5\text{ A})(0.21\text{ m})\times(48\text{ mT}) = 21.4\text{ mN}$ and direction 45° between the positive x and y axes.

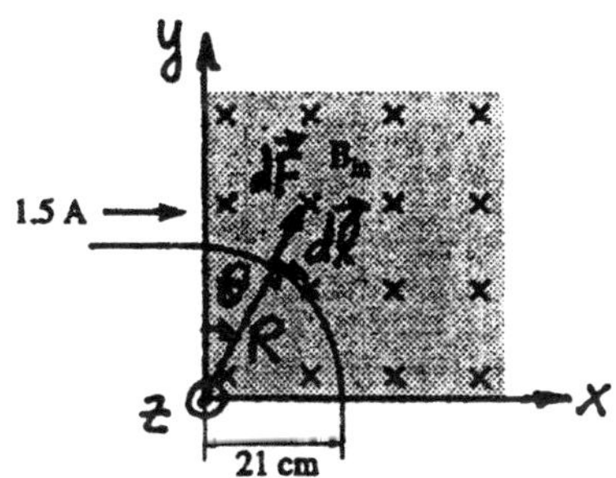

FIGURE 29-43 Problem 41 Solution.

Problem

43. Apply Equation 29-8 to a closed current loop of arbitrary shape in a *uniform* magnetic field, and show that the net force on the loop is zero. *Hint:* Both I and $\mathbf{B}$ are constant as you go around the loop, so you can take them out of the integral. What is the remaining vector integral?

Solution

If both I and $\mathbf{B}$ are constants, Equation 29-8, for a closed loop, may be written as $\mathbf{F} = \oint I d\boldsymbol{\ell}\times\mathbf{B} = I(\oint d\boldsymbol{\ell})\times\mathbf{B}$. But the integral is the sum of vectors, $d\boldsymbol{\ell}$, beginning and ending at the same point, hence $\oint d\boldsymbol{\ell} = 0$. (This integral relation is a special case of Equation 8-1 for a constant $\mathbf{F}$.)

Problem

45. The probe in a Hall-effect magnetometer uses a semiconductor doped to a charge-carrier density of $7.5\times10^{20}\text{ m}^{-3}$. The probe measures 0.35 mm thick in the direction of the magnetic field being measured, and carries a 2.5-mA current perpendicular to the field. If its Hall potential is 4.5 mV, what is the magnetic field strength?

Solution

If we assume the charge-carriers are of one type, with charge of magnitude e, then Equation 29-7 and the given data require $B = nqV_Ht/I = (7.5\times10^{20}\text{ m}^{-3})\times(1.6\times10^{-19}\text{ C})(4.5\text{ V})(0.35\text{ mm})/(2.5\text{ mA}) = 75.6\text{ mT}$.

Section 29-5: A Current Loop in a Magnetic Field

Problem

47. A single-turn square wire loop 5.0 cm on a side carries a 450-mA current. (a) What is the magnetic moment of the loop? (b) If the loop is in a uniform 1.4-T magnetic field with its dipole moment vector at 40° to the field direction, what is the magnitude of the torque it experiences?

Solution

(a) Equation 29-10 for the magnetic moment of a loop gives $\mu = NIA = (1)(450\text{ mA})(5\text{ cm})^2 = 1.13\times10^{-3}\text{ A}\cdot\text{m}^2$. (b) Equation 29-11 gives the torque on a magnetic dipole moment in a uniform magnetic field, $\tau = |\boldsymbol{\mu}\times\mathbf{B}| = \mu B\sin\theta = (1.13\times10^{-3}\text{ A}\cdot\text{m}^2)(1.4\text{ T})\times\sin 40^\circ = 1.01\times10^{-3}\text{ N}\cdot\text{m}$.

Problem

49. A bar magnet experiences a 12-mN·m torque when it is oriented at 55° to a 100-mT magnetic field. What is the magnitude of its magnetic dipole moment?

Solution

Equation 29-11, solved for the magnitude of the dipole moment, gives $\mu = \tau/B\sin\theta = (12\times10^{-3}\text{ N}\cdot\text{m})\div(0.1\text{ T})\sin 55^\circ = 0.146\text{ A}\cdot\text{m}^2$.

Problem

51. A simple electric motor like that of Fig. 29-36 consists of a 100-turn coil 3.0 cm in diameter, mounted between the poles of a magnet that produces a 0.12-T field. When a 5.0-A current flows in the coil, what are (a) its magnetic dipole moment and (b) the maximum torque developed by the motor?

Solution

(a) From Equation 29-10, the magnetic moment of the coil has magnitude $\mu = NIA = 100(5\ \text{A})\frac{1}{4}\pi(0.03\ \text{m})^2 = 0.353\ \text{A}\cdot\text{m}^2$. The direction of μ, determined from the right-hand rule (see Fig. 29-33), rotates with the coil. (b) The maximum torque (from Equation 29-11, with $\sin\theta = 1$) is $\tau_{\text{max}} = \mu B = (0.353\ \text{A}\cdot\text{m}^2)(0.12\ \text{T}) = 4.24\times10^{-2}\ \text{N}\cdot\text{m}$.

Problem

53. Nuclear magnetic resonance (NMR) is a technique for analyzing chemical structures and is also the basis of magnetic resonance imaging used for medical diagnosis. The NMR technique relies on sensitive measurements of the energy needed to flip atomic nuclei upside-down in a given magnetic field. In an NMR apparatus with a 7.0-T magnetic field, how much energy is needed to flip a proton ($\mu = 1.41\times10^{-26}\ \text{A}\cdot\text{m}^2$) from parallel to antiparallel to the field?

Solution

From Equation 29-12, the energy required to reverse the orientation of a proton's magnetic moment from parallel to antiparallel to the applied magnetic field is $\Delta U = 2\mu B = 2(1.41\times10^{-26}\ \text{A}\cdot\text{m}^2)(7.0\ \text{T}) = 1.97\times10^{-25}\ \text{J} = 1.23\times10^{-6}\ \text{eV}$. (This amount of energy is characteristic of radio waves of frequency 298 MHz, see Chapter 39.)

Paired Problems

Problem

55. Find the magnetic force on an electron moving with velocity $\mathbf{v} = 8.6\times10^5\hat{\imath} - 4.1\times10^5\hat{\jmath}$ m/s in a magnetic field $\mathbf{B} = 0.18\hat{\jmath} + 0.64\hat{\mathbf{k}}$ T.

Solution

From Equation 29-1a, $\mathbf{F} = -e\mathbf{v}\times\mathbf{B} = -(1.6\times10^{-19}\ \text{C})(8.6\hat{\imath} - 4.1\hat{\jmath})(10^5\ \text{m/s})\times(0.18\hat{\jmath} + 0.64\hat{\mathbf{k}})\ \text{T} = (16\ \text{fN})(4.1\times0.64\hat{\imath} + 8.6\times0.64\hat{\jmath} - 8.6\times0.18\hat{\mathbf{k}}) = (42.0\hat{\imath} + 88.1\hat{\jmath} - 24.8\hat{\mathbf{k}})\ \text{fN}$. (If necessary, review the cross product of unit vectors; see Fig. 13-9 and use the right hand rule.)

Problem

57. Proponents of space-based particle-beam weapons have to confront the effect of Earth's magnetic field on their beams. If a beam of protons with kinetic energy 100 MeV is aimed in a straight line perpendicular to Earth's magnetic field in a region where the field strength is 48 μT, what will be the radius of the protons' circular path?

Solution

It is simplest to use the result of Problem 22, in atomic units, to find the radius (see solution to Problem 19), $r = \sqrt{2Km}/300qB$, where r is in meters, K is in MeV, m in MeV/c^2, q in units of e, and B in teslas. Then $r = \sqrt{2(100)(938)}/300(0.48\times10^{-4}) = 30.1$ km. (The mass of a proton in atomic units is approximately 938 MeV/c^2.)

Problem

59. A 170-mT magnetic field points into the page, confined to a square region as shown in Fig. 29-44. A square conducting loop 32 cm on a side carrying a 5.0-A current in the clockwise sense extends partly into the field region, as shown. Find the magnetic force on the loop.

Solution

The forces on the upper and lower horizontal parts of the loop in the field region cancel one-another, so the net force is just due to the magnetic force on the right vertical side, which is directed to the right in Fig. 29-44, with magnitude $I\ell B = (5\ \text{A})(0.32\ \text{m})\times(170\ \text{mT}) = 0.272$ N. (See Equation 29-6.)

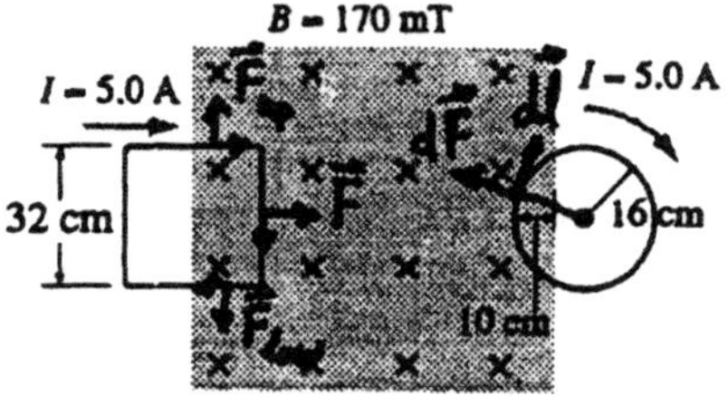

FIGURE 29-44 Problem 59 Solution.

Problem

61. An old-fashioned analog meter uses a wire coil in a magnetic field to deflect the meter needle. If the coil is 2.0 cm in diameter and consists of 500 turns of wire, what should be the magnetic field strength if the maximum torque is to be 1.6 μN·m when the current in the coil is 1.0 mA?

Solution

The maximum torque on a flat coil, with magnetic moment $\mu = N\pi R^2 I$, is $\tau_{\text{max}} = \mu B$ (see Equations 29-10 and 11). Thus, $B = \tau_{\text{max}}/NI\pi R^2 = (1.6\ \mu\text{N}\cdot\text{m})/(500)(1\ \text{mA})\pi(1\ \text{cm})^2 = 102$ G.

Supplementary Problems

Problem

63. Electrons in a TV picture tube are accelerated through a 30-kV potential difference and head straight for the center of the tube, 40 cm away. If the electrons are moving at right angles to Earth's 0.50-G magnetic field, by how much do they miss the screen's exact center?

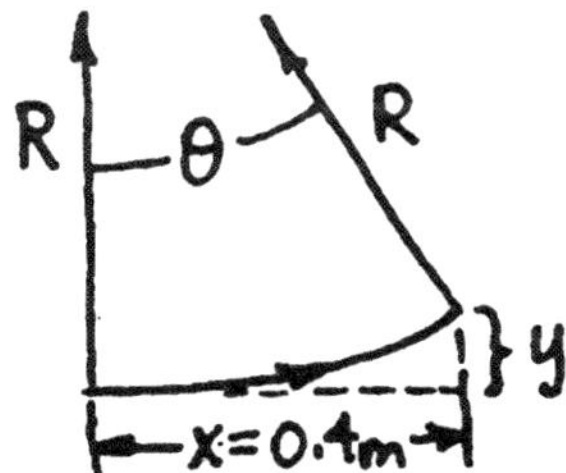

Problem 63 Solution.

Solution

The electrons suffer the maximum deflection when moving perpendicularly to the magnetic field. They are bent into a circle of radius

$$R = \sqrt{2Km}/eB$$

$$= \frac{\sqrt{2(3\times10^4\text{ eV})(1.6\times10^{-19}\text{ J/eV})(9.11\times10^{-31}\text{ kg})}}{(1.6\times10^{-19}\text{ C})(5\times10^{-5}\text{ T})}$$

$$= 11.7\text{ m}.$$

Geometry gives the maximum deflection from screen center: $y = R(1 - \cos\theta)$, where $\sin\theta = x/R$. Numerically, $y = (11.7\text{ m})\{1 - \cos[\sin^{-1}(0.4/11.7)]\} =$ 6.85 mm.

Problem

65. A conducting bar with mass 15.0 g and length 22.0 cm is suspended from a spring in a region where a 0.350-T magnetic field points into the page, as shown in Fig. 29-45. With no current in the bar, the spring length is 26.0 cm. The bar is supplied with current from outside the field region, using wires of negligible mass. When a 2.00-A current flows from left to right in the bar, it rises 1.2 cm from its equilibrium position. Find (a) the spring constant and (b) the unstretched length of the spring.

Solution

In equilibrium, the vertical forces on the bar must sum to zero. These include the weight of the bar, $-mg$ (negative downward), the upward spring force

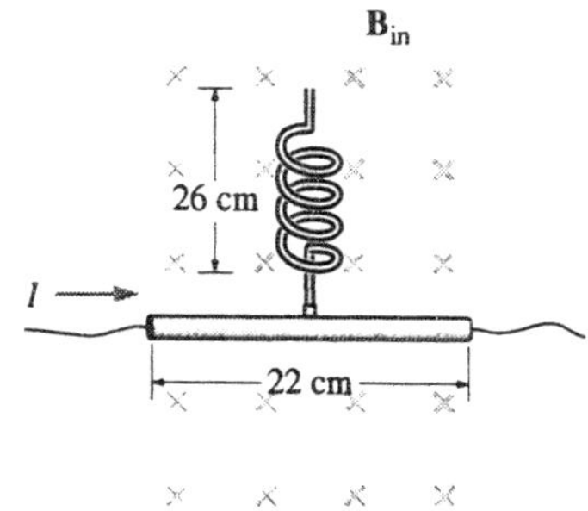

FIGURE 29-45 Problem 65.

$k\,\Delta\ell = k(\ell - \ell_0)$ (where ℓ is the length and ℓ_0 the unstretched length of the spring), and when current flows from left to right in the bar, an upward magnetic force of $F_B = ILB$ (L is the length of the bar). The conditions stated require that $k(26.0\text{ cm} - \ell_0) - mg = 0$, and $k(24.8\text{ cm} - \ell_0) + ILB - mg = 0$ or $k(26\text{ cm} - \ell_0) = mg$ and $k(26\text{ cm} - \ell_0 - 1.2\text{ cm}) = mg - ILB$. These equations can be solved for k and ℓ_0 (subtract to eliminate ℓ_0, and divide to eliminate k) with the result $k = ILB/(1.2\text{ cm}) = (2\text{ A})(22\text{ cm})\times(0.35\text{ T})/(1.2\text{ cm}) = 12.8\text{ N/m}$, and $\ell_0 = 26\text{ cm} - (mg/ILB)(1.2\text{ cm}) = 26\text{ cm} - (0.015\times9.8\text{ N})\div(12.8\text{ N/m}) = 24.9\text{ cm}$.

Problem

67. A solid disk of mass M and thickness d sits on an incline, as shown in Fig. 29-46. A loop of wire is wrapped around the disk, running along a diameter and oriented so the loop is parallel to the incline. A uniform magnetic field **B** points

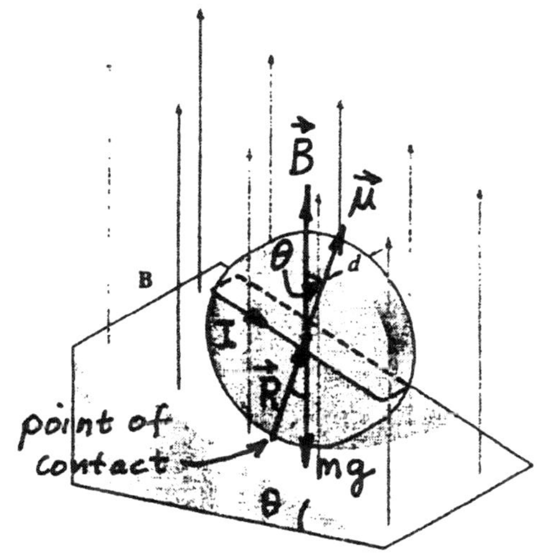

FIGURE 29-46 Problem 67.

vertically upward. Find an expression for the current I in the loop that will keep it from rolling down the incline.

Solution

The torque of gravity causing the disk to roll down the incline (about the point of contact) is $\tau_{grav} = mgR\times \sin(180°-\theta) = mgR\sin\theta$ into the page in Fig. 29-46. The magnetic torque on the magnetic moment of the loop, which would cancel this, is $\tau_{mag} = \mu B\sin\theta$ out of the page for the current shown, where $\mu = IA$ and $A = 2Rd$ is the area of the loop. (The magnetic torque is due to a couple and is the same about any point.) In equilibrium, $mg\sin\theta = I(2Rd)B\sin\theta$, or $I = mg/2Bd$. (We assume that static friction between the disk and the incline is sufficient to satisfy the other conditions for equilibrium.)

Problem

69. A 10-turn wire loop measuring 8.0 cm by 16 cm carrying 2.0 A lies in a horizontal plane but is free to rotate about the axis shown in Fig. 29-47. A 50-g mass hangs from one side of the loop, and a uniform magnetic field points horizontally, as shown. What magnetic field strength is required to hold the loop in its horizontal position?

Solution

For the direction of current shown in Fig. 29-47, the magnetic moment of the loop is downward, and the magnetic torque $\boldsymbol{\mu}\times\mathbf{B}$ is along the axis, out of the page. The gravitational torque $\mathbf{r}\times m\mathbf{g}$ is along the axis, into the page. The two torques cancel when $\mu B = mgr$, or $B = mgr/NIA = (0.05\text{ kg})(9.8\text{ m/s}^2)\times(0.04\text{ m})/(10)(2.0\text{ A})(0.8\times0.16\text{ m}^2) = 76.6\text{ mT}$.

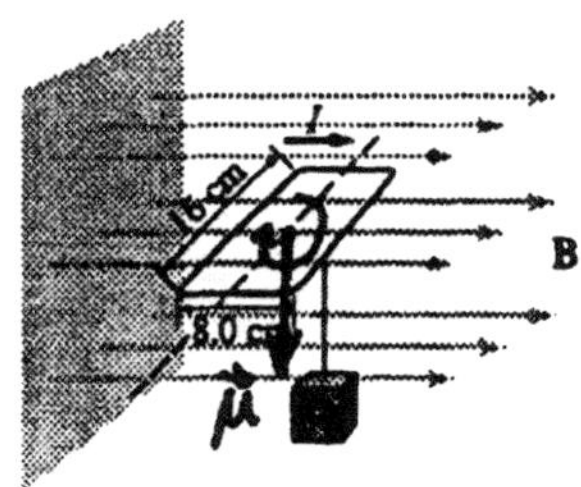

FIGURE 29-47 Problem 69 Solution.

Problem

71. A circular wire loop of mass m and radius R carries a current I. The loop is hanging horizontally below a cylindrical bar magnet, suspended by the magnetic force, as shown in Fig. 29-49. If the field lines crossing the loop make an angle θ with the vertical, show that the strength of the magnet's field at the loop's position is $B = mg/2\pi RI\sin\theta$.

Solution

We assume that the magnetic field has axial symmetry, with vertical axis through the center of the loop's horizontal plane area. The magnetic field lines intersect the loop at right angles, and if the current circulates clockwise (as seen from the magnet), the magnetic force on an element $I\,d\ell$ has an upward component $dF_y = dF\sin\theta = I\,d\ell\,B\sin\theta$. This is the same at any position on the loop, so the net upward force is $F_y = \int IB\sin\theta\,d\ell = 2\pi RIB\sin\theta$. (The horizontal forces on diametrically opposite elements of loop cancel, so $F_x = 0$.) When the loop is suspended, the upward magnetic force balances the loop's weight, so $F_y = mg$, or $B = mg/2\pi RI\sin\theta$.

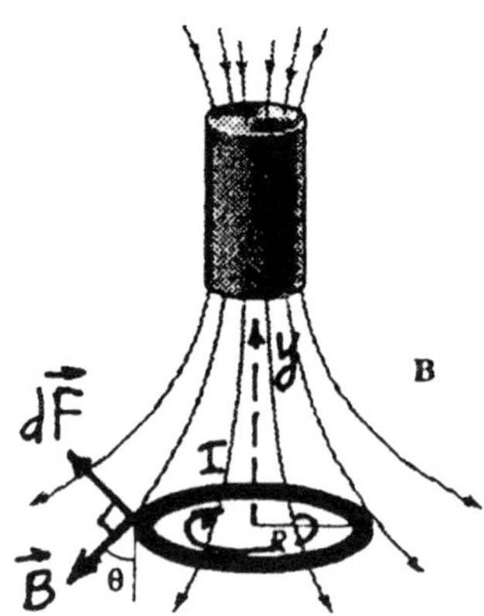

FIGURE 29-49 Problem 71 Solution.

Problem

73. Early models pictured the electron in a hydrogen atom as being in a circular orbit of radius 5.29×10^{-11} m about the stationary proton, held in orbit by the electric force. Find the magnetic dipole moment of such an atom. This quantity is called the *Bohr magneton* and is typical of atomic-sized magnetic moments. *Hint:* The full electron charge passes any given point in the orbit once per orbital period. Use this fact to calculate the average current.

Solution

One electronic charge passes a given point on the orbit every period of revolution, so the magnitude of the average current corresponding to the electron's orbital motion is $I = \Delta q/\Delta t = e/(2\pi r/v)$. (The current circulates opposite to the orbital motion, since the electron is negatively charged.) In the simplest version of the Bohr model for the hydrogen atom, the electron moves in a circular orbit, around a fixed proton, under the influence of the Coulomb force, so that $mv^2/r = ke^2/r^2$, or $v/r = \sqrt{ke^2/mr^3}$. Thus, $I = (e/2\pi)(v/r) = (e^2/2\pi)\sqrt{k/mr^3}$. The magnetic dipole moment associated with this orbital atomic current (called a Bohr magneton) has magnitude $\mu_B = I\pi r^2 = \frac{1}{2}e^2\sqrt{kr/m} \approx \frac{1}{2}(1.6\times10^{-19}\text{ C})^2\times \sqrt{(9\times10^9\text{ N}\cdot\text{m}^2/\text{C}^2)(5.29\times10^{-11}\text{ m})/(9.11\times10^{-31}\text{ kg})} \approx 9.25\times10^{-24}\text{ A}\cdot\text{m}^2$, and is typical of the size of atomic magnetic dipole moments in general.

CHAPTER 30 SOURCES OF THE MAGNETIC FIELD

ActivPhysics can help with these problems: Activities 13.1–13.3, 13.5

Section 30-1: The Biot-Savart Law

Problem

1. A wire carries 15 A. You form the wire into a single-turn circular loop with magnetic field 80 μT at the loop center. What is the loop radius?

Solution

Equation 30-3, with $x = 0$, gives the magnetic field at the center of a circular loop, $B = \mu_0 I/2a$ (with direction along the axis of the loop, consistent with the sense of circulation of the current and the right-hand rule). Thus, the radius is $a = \mu_0 I/2B = (2\pi\times10^{-7}\ \text{N/A}^2)(15\ \text{A})/(80\ \mu\text{T}) = 11.8$ cm.

Problem

3. A 2.2-m-long wire carrying 3.5 A is wound into a tight, loop-shaped coil 5.0 cm in diameter. What is the magnetic field at its center?

Solution

The magnetic field at the center of each of the tightly-wound turns is the same (Equation 30-3 with $x = 0$), so the net field is $B = N(\mu_0 I/2a)$, where N is the number of turns. If the wire has length L and the coil has diameter D, then $N = L/\pi D$ and $B = (L/\pi D)(\mu_0 I/D) = \mu_0 IL/\pi D^2 = (4\times10^{-7}\ \text{N/A}^2)\times (3.5\ \text{A})(2.2\ \text{m})/(5\ \text{cm})^2 = 1.23$ mT.

Problem

5. Suppose Earth's magnetic field arose from a single loop of current at the outer edge of the planet's liquid core (core radius 3000 km), concentric with Earth's center. What current would be necessary to give the observed field strength of 62 μT at the north pole? (The currents responsible for Earth's field are more complicated than this problem suggests.)

Solution

The north pole is on the axis of the hypothetical circular current loop, with radius $a = 3.0$ Mm, at a distance $x = R_E = 6.37$ Mm from the earth's center. Equation 30-3 gives a current of $I = 2B(x^2 + a^2)^{3/2} \div \mu_0 a^2 = 2(62\ \mu\text{T})[(6.37\ \text{Mm})^2 + (3.0\ \text{Mm})^2]^{3/2} \div (4\pi\times10^{-7}\ \text{T}\cdot\text{m/A})(3.0\ \text{Mm})^2 = 3.83$ GA.

Problem

7. A single-turn current loop carrying 25 A produces a magnetic field of 3.5 nT at a point on its axis 50 cm from the loop center. What is the loop area, assuming the loop diameter is much less than 50 cm?

Solution

If the radius of the loop is assumed to be much smaller than the distance to the field point ($a \ll x = 50$ cm), then Equation 30-4 for the field on the axis of a magnetic dipole can be used to find $\mu = 2\pi x^3 B/\mu_0$. The magnetic moment of a single-turn loop is $\mu = IA$, therefore $A = \mu/I = 2\pi x^3 B/\mu_0 I = (3.5\ \text{nT})(50\ \text{cm})^3 \div (2\times10^{-7}\ \text{N/A}^2)(25\ \text{A}) = 0.875\ \text{cm}^2$.

Problem

9. You have a spool of thin wire that can handle a maximum current of 0.50 A. If you wind the wire into a loop-like coil 20 cm in diameter, how many turns should the coil have if the magnetic field at its center is to be 2.3 mT at this maximum current?

Solution

Equation 30-3 can be modified for N turns of wire (as in the solution to Problem 3), so at the center of a flat circular coil, $B = N\mu_0 I/2a$. Thus, $N = 2aB/\mu_0 I = (0.2\ \text{m})(2.3\ \text{mT})/(4\pi\times10^{-7}\ \text{N/A}^2)(0.5\ \text{A}) = 732$.

Problem

11. Two long, parallel wires are 6.0 cm apart. One carries 5.0 A and the other 10 A, with both currents in the same direction. Where on a line perpendicular to both wires is the magnetic field zero?

Solution

The magnetic fields from each parallel wire are in opposite directions at a point between the wires, with magnitudes given by Equation 30-5 (or Equation 30-8). If d is the distance between the wires and x the distance of the field point from the first wire, then $B = B_1 - B_2 = (\mu_0/2\pi)[(I_1/x) - I_2/(d-x)]$. Evidently, $B = 0$ when $I_1/x = I_2/(d-x)$, or $x = d/(1 + I_2/I_1)$. For the case $I_2 = 2I_1$ and $d = 6$ cm, this gives a null field at $x = 2$ cm from the first (smaller current) wire.

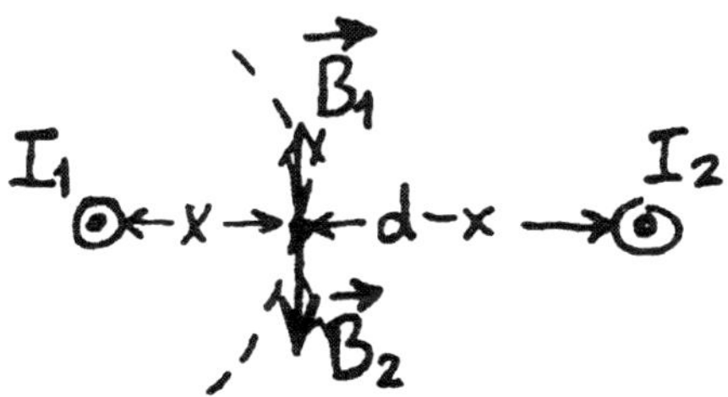

Problem 11 Solution.

Problem

13. A power line carries a 500-A current toward magnetic north and is suspended 10 m above the ground. The horizontal component of Earth's magnetic field at the power line's latitude is 0.24 G. If a magnetic compass is placed on the ground directly below the power line, in what direction will it point?

Solution

A compass needle (small dipole magnet) is free to rotate in a horizontal plane until it is aligned with the direction of the total horizontal magnetic field (see Equation 29-11). A long, straight wire (the power line) carrying a current of 500 A parallel to the ground, in the direction of magnetic north, produces a magnetic field, at a distance 10 m below, to the west, with magnitude given by Equation 30-8 (see diagram): $B_y = \mu_0 I/2\pi r = (2\times10^{-7}\text{ N/A}^2)(500\text{ A})/10\text{ m} = 10^{-5}\text{ T} = 0.1\text{ G}$. The horizontal component of the Earth's magnetic field is $B_x = 0.24$ G, so the compass needle will point $\theta = \tan^{-1} B_y/B_x = \tan^{-1}(0.1/0.24) = 22.6°$ west of magnetic north.

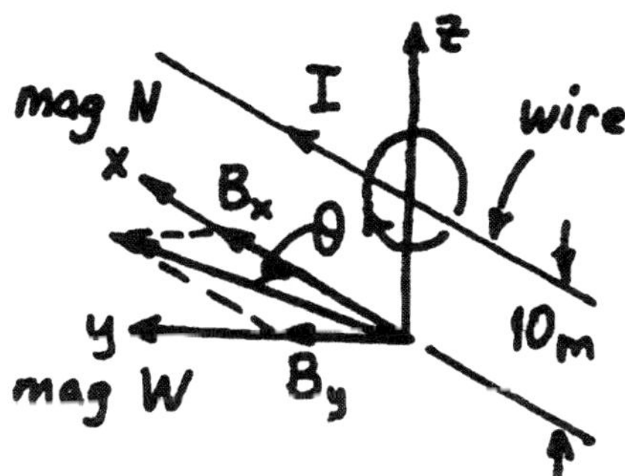

Problem 13 Solution.

Problem

15. Part of a long wire is bent into a semicircle of radius a, as shown in Fig. 30-50. A current I flows in the direction shown. Use the Biot-Savart law to find the magnetic field at the center of the semicircle (point P).

Solution

The Biot-Savart law (Equation 30-2) written in a coordinate system with origin at P, gives $\mathbf{B}(P) = (\mu_0 I/4\pi)\int_{\text{wire}} d\boldsymbol{\ell}\times\hat{\mathbf{r}}/r^2$, where $\hat{\mathbf{r}}$ is a unit vector from an element $d\boldsymbol{\ell}$ on the wire to the field point P. On the straight segments to the left and right of the semicircle, $d\boldsymbol{\ell}$ is parallel to $\hat{\mathbf{r}}$ or $-\hat{\mathbf{r}}$, respectively, so $d\boldsymbol{\ell}\times\hat{\mathbf{r}} = 0$. On the semicircle, $d\boldsymbol{\ell}$ is perpendicular to $\hat{\mathbf{r}}$ and the radius is constant, $r = a$. Thus,

$$B(P) = \frac{\mu_0 I}{4\pi}\int_{\text{semicircle}} \frac{d\ell}{a^2} = \frac{\mu_0 I}{4\pi}\cdot\frac{\pi a}{a^2} = \frac{\mu_0 I}{4a}.$$

The direction of $\mathbf{B}(P)$, from the cross product, is into the page.

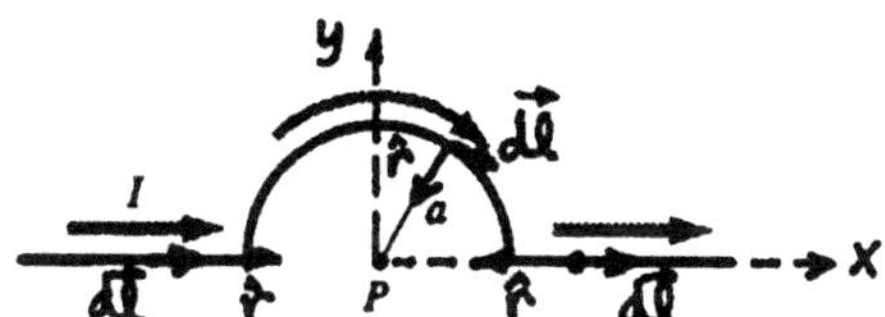

FIGURE 30-50 Problem 15 Solution.

Problem

17. Figure 30-51 shows a conducting loop formed from concentric semicircles of radii a and b. If the loop carries a current I as shown, find the magnetic field at point P, the common center.

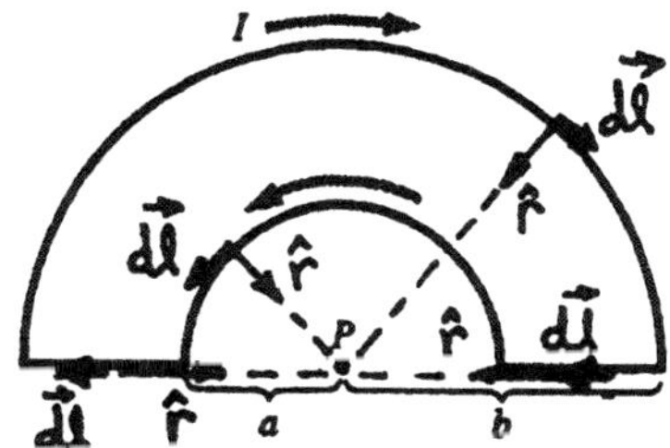

FIGURE 30-51 Problem 17 Solution.

Solution

The Biot-Savart law gives $\mathbf{B}(P) = (\mu_0/4\pi)\times\int_{\text{loop}} I d\boldsymbol{\ell}\times\hat{\mathbf{r}}/r^2$. $I d\boldsymbol{\ell}\times\hat{\mathbf{r}}/r^2$ on the inner semicircle has magnitude $Id\ell/a^2$ and direction out of the page, while on the outer semicircle, the magnitude is $Id\ell/b^2$ and the direction is into the page. On the straight segments, $d\boldsymbol{\ell} \times \hat{\mathbf{r}} = 0$, so the total field at P is $(\mu_0/4\pi)[(I\pi a/a^2) - (I\pi b/b^2)] = \mu_0 I(b-a)/4ab$ out of the page. (Note: the length of each semicircle is $\int d\ell = \pi r$.)

Section 30-2: The Magnetic Force Between Two Conductors

Problem

19. It would take a rather large apparatus to implement the definition of the ampere given at the end of Section 30-2. Suppose you wanted to use a smaller apparatus, with wires 50 cm long separated by 2.0 cm. What force would correspond to a current of 1 A?

Solution

Since 50 cm $= \ell \gg d = 2$ cm, the wires can be considered approximately infinitely long and Equation 30-6 gives $F = \mu_0 I_1 I_2 \ell/2\pi d = (2\times10^{-7}\ \text{N/A}^2)\times(1\ \text{A})^2(50\ \text{cm})/(2\ \text{cm}) = 5\ \mu\text{N}$ (about half the weight of a raindrop, see Table 1-4).

Problem

21. The structure shown in Fig. 30-52 is made from conducting rods. The upper horizontal rod is free to slide vertically on the uprights, while maintaining electrical contact with them. The upper rod has mass 22 g and length 95 cm. A battery connected across the insulating gap at the bottom of the left-hand upright drives a 66-A current through the structure. At what height h will the upper wire be in equilibrium?

Solution

If h is small compared to the length of the rods, we can use Equation 30-6 for the repulsive magnetic force between the horizontal rods (upward on the top rod) $F = \mu_0 I^2 \ell/2\pi h$. The rod is in equilibrium when this equals its weight, $F = mg$, hence $h = \mu_0 I^2 \ell/2\pi mg = (2\times10^{-7}\ \text{N/A}^2)(66\ \text{A})^2(0.95\ \text{m})/(0.022\times9.8\ \text{N}) = 3.84$ mm. (This is indeed small compared to 95 cm, as assumed.)

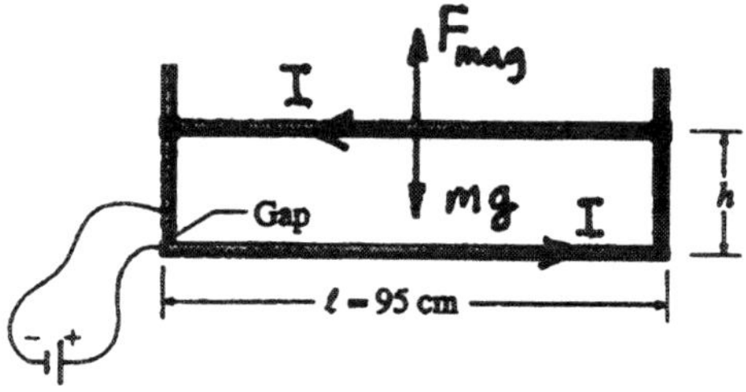

FIGURE 30-52 Problem 21 Solution.

Problem

23. The wires in Fig. 30-49 carry 2.5-A currents in the directions indicated. Find the net force per unit length on the wire at lower left.

Solution

The force per unit length between any pair of wires has magnitude given by Equation 30-6, and is attractive for parallel and repulsive for antiparallel currents. For the bottom pair, for example, the force per unit length on the lower left wire is $F/\ell = (\mu_0 I^2/2\pi d)$ to the right, where d is the side of the square. The other forces on this wire are (F/ℓ) down and $(F/\ell)/\sqrt{2}$ diagonally down and left, as shown. The resultant force is $(F/\ell)\hat{\imath} + (F/\ell)(-\hat{\jmath}) + (F/\ell)(-\hat{\imath}\cos 45^\circ - \hat{\jmath}\sin 45^\circ)/\sqrt{2} = (F/\ell)(\frac{1}{2}\hat{\imath} - \frac{3}{2}\hat{\jmath})$. This has magnitude $\frac{1}{2}\sqrt{10}(F/\ell) = \frac{1}{2}\sqrt{10}(2\times10^{-7}\ \text{N/A}^2)(25\ \text{A})^2/(0.15\ \text{m}) = 13.2\ \mu\text{N/m}$, and direction $\theta = \tan^{-1}(-3) = -71.6^\circ$ below x axis shown (lower right quadrant).

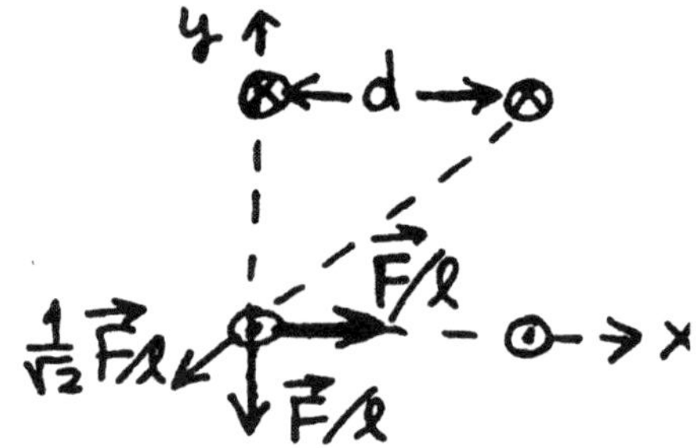

Problem 23 Solution.

Problem

25. A solenoid 10 cm in diameter is made with 2.1-mm-diameter copper wire wound so tightly that adjacent turns touch, separated only by enamel insulation of negligible thickness. The solenoid carries a 28-A current. In the long, straight wire approximation, what is the net force between two adjacent turns of the solenoid?

Solution

Equation 30-6 provides an approximate value of the force per unit length between adjacent turns of $F/\ell = \mu_0 I^2/2\pi d = (2\times10^{-7}\ \text{N/A}^2)(28\ \text{A})^2 \div (2.1\ \text{mm}) = 74.7$ mN/m. The length of one turn is $2\pi r = 10\pi$ cm $= 0.314$ m, so the force on one turn is approximately $(74.7\ \text{mN/m})(0.314\ \text{m}) = 23.5$ mN.

Section 30-3: Ampère's Law

Problem

27. The line integral of the magnetic field on a closed path surrounding a wire has the value 8.8 μT·m. What is the current in the wire?

Solution

If the only current encircled by the path is that flowing in the wire, Ampère's law (Equation 30-7) gives 8.8 μT·m $= \oint \mathbf{B}\cdot d\boldsymbol{\ell} = \mu_0 I_{\text{wire}}$, or $I_{\text{wire}} =$ (8.8 μT·m)/($4\pi\times10^{-7}$ N/A^2) = 7.00 A.

Problem

29. The magnetic field shown in Fig. 30-56 has uniform magnitude 75 μT, but its direction reverses abruptly. How much current is encircled by the rectangular loop shown?

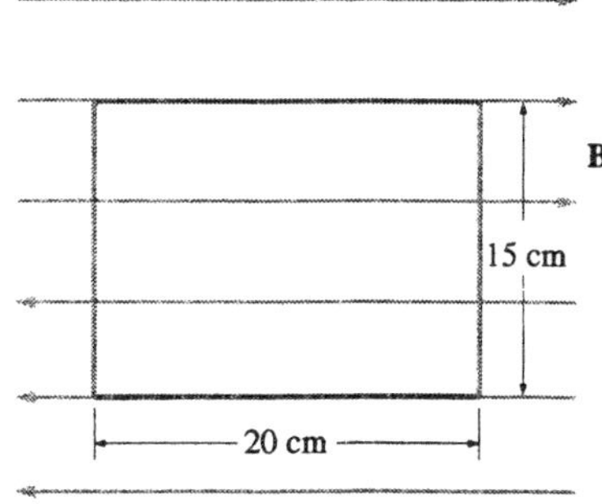

FIGURE 30-56 Problem 29.

Solution

Ampère's law applied to the loop shown in Fig. 30-56 (going clockwise) gives $\oint \mathbf{B}\cdot d\boldsymbol{\ell} = 2B\ell = 2(75\ \mu\text{T})\times$ $(0.2\ \text{m}) = \mu_0 I_{\text{encircled}} = (0.4\pi\mu\text{T}\cdot\text{m/A})I_{\text{encircled}}$ (since the sides of the loop perpendicular to **B** give no contribution to the line integral). Thus, $I_{\text{encircled}} = (75/\pi)$ A = 23.9 A. As explained in the text, the current flows along the boundary surface between the regions of oppositely directed **B**, positive into the page in Fig. 30-56, for clockwise circulation around the loop.

Problem

31. Figure 30-58 shows a magnetic field pointing in the x direction. Its strength, however, varies with position in the y direction. At the top and bottom of the rectangular loop shown the field strengths are 3.4 μT and 1.2 μT, respectively. How much current flows through the area encircled by the loop?

Solution

Equation 30-7 applied to the loop shown (going around in the direction of **B** at the top, i.e., clockwise) gives $\oint \mathbf{B}\cdot d\boldsymbol{\ell} = B_{\text{top}}\ell - B_{\text{bot}}\ell = (3.4-1.2)\ \mu\text{T}(7\ \text{cm}) =$ $\mu_0 I_{\text{encircled}}$, so $I_{\text{encircled}} = (2.2\ \mu\text{T})(0.07\ \text{m})\div$ $(0.4\pi\ \mu\text{T}\cdot\text{m/A}) = 123$ mA (positive current into the page in Fig. 30-58). (Note: $\mathbf{B}\cdot d\boldsymbol{\ell} = 0$ on the 4 cm sides of the amperian loop and a right-hand screw turned clockwise advances into the page.)

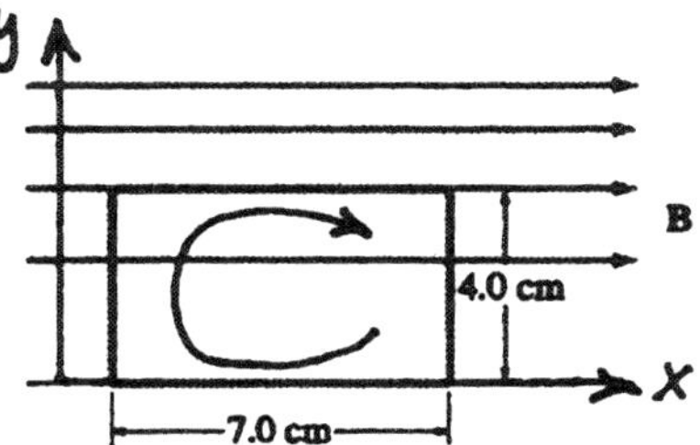

FIGURE 30-58 Problem 31 Solution.

Section 30-4: Using Ampère's Law

Problem

33. A solid wire 2.1 mm in diameter carries a 10-A current with uniform current density. What is the magnetic field strength (a) at the axis of the wire, (b) 0.20 mm from the axis, (c) at the surface of the wire, and (d) 4.0 mm from the wire axis?

Solution

The magnetic field strength is given by Equation 30-9 inside the wire ($r \le R$) and Equation 30-8 outside ($r \ge R$) as shown in Fig. 30-24. (a) For $r = 0, B = 0$. (b) For $r = 0.2$ mm $< R = \frac{1}{2}\times2.1$ mm $= 1.05$ mm, $B = \mu_0 I r/2\pi R^2 = (2\times10^{-7}\ \text{N/A}^2)(10\ \text{A})\times$ $(0.2\ \text{mm})/(1.05\ \text{mm})^2 = 3.63$ G. (c) For $r = R$, $B =$ $\mu_0 I/2\pi R = (2\times10^{-7}\ \text{N/A}^2)(10\ \text{A})/(1.05\ \text{mm}) =$ 19.0 G. (d) For $r = 4$ mm $> R$, $B = \mu_0 I/2\pi r =$ $(2\times10^{-7}\ \text{N/A}^2)(10\ \text{A})/(4\ \text{mm}) = 5$ G.

Problem

35. A long conducting rod of radius R carries a nonuniform current density given by $J = J_0 r/R$, where J_0 is constant and r is the radial distance from the rod's axis. Find expressions for the magnetic field strength (a) inside and (b) outside the rod.

Solution

The magnetic field from a long conducting rod is approximately cylindrically symmetric, as discussed in Section 30-4, so (b) the field outside ($r \ge R$) is given by Equation 30-8, and has direction circling the rod according to the right-hand rule. The total current can be related to the current density by integrating over the cross-sectional area of the rod, $I = \int \mathbf{J}\cdot d\mathbf{A} = \int_0^R J_0(r/R)2\pi r\, dr = 2\pi J_0 R^2/3$. Here, area elements were chosen to be circular rings of radius r, thickness dr, and area $dA = 2\pi r\, dr$. Equation 30-8 can then be written as $B = \mu_0 I/2\pi r = \mu_0 J_0 R^2/3r$, for $r \geq R$. (a) Inside the rod, Ampère's law can be used to find the field, as in Example 30-4, by integrating the current density over a smaller cross-sectional area, corresponding to $I_{\text{encircled}}$ for an amperian loop with $r \leq R$. Then $I_{\text{encircled}} = \int_0^r J_0(r/R)2\pi r\, dr = 2\pi J_0 r^3/3R$, and Ampère's law gives $2\pi r B = \mu_0 I_{\text{encircled}}$, or $B = \mu_0 J_0 r^2/3R$, for $r \leq R$.

Problem

37. Typically, cylindrical wires made from yttrium-barium-copper-oxide superconductor can carry a maximum current density of 6.0 MA/m^2 at a temperature of 77 K, as long as the magnetic field at the conductor surface does not exceed 10 mT. Suppose such a wire is to carry the maximum current density. (a) At what wire diameter would the surface magnetic field equal the 10-mT limit? (b) Is this a maximum or minimum value for the diameter if the field is not to exceed the limit? (c) What current would a wire with this diameter carry?

Solution

The magnetic field strength at the surface of a wire with axial symmetry is $B = \mu_0 I/2\pi R$ (see Problem 34). If the current density in the wire is uniform over its circular cross-section, $I = J\pi R^2$ and $B = \frac{1}{2}\mu_0 JR$. (a) and (b) If $J = J_{\text{max}}$, then $B \leq B_{\text{max}}$ implies $2R \leq 4B_{\text{max}}/\mu_0 J_{\text{max}} = (10 \text{ mT}) \div (0.1\pi\ \mu\text{T}\cdot\text{m/A})/(6 \text{ MA/m}^2) = 5.31$ mm, which is the maximum diameter. (c) With the diameter above, the total current is $I = J_{\text{max}}\pi R^2 = (6 \text{ MA/m}^2)\pi\times(5.31 \text{ mm}/2)^2 = 133$ A, not excessive for a superconductor.

Problem

39. Two large, flat conducting plates lie parallel to the x-y plane. They carry equal currents, one in the $+x$ and the other in the $-x$ direction. In each plate the current per meter of width in the y direction is J_s. Find the magnetic field strength (a) between and (b) outside the plates.

Solution

The total field is the superposition of fields due to two (approximately infinite) flat parallel current sheets. Equation 30-10 and Fig. 30-25 give the magnitude and direction of the individual fields. (a) Between the plates, both fields are in the negative y direction, thus, $\mathbf{B}_{\text{btw}} = -\frac{1}{2}\mu_0 J_{s,1}\hat{\mathbf{j}} - \frac{1}{2}\mu_0 J_{s,2}\hat{\mathbf{j}} = -\mu_0 J_s\hat{\mathbf{j}}$. (b) Outside the plates, the fields are in opposite directions, and thus cancel, $\mathbf{B}_{\text{out}} = 0$.

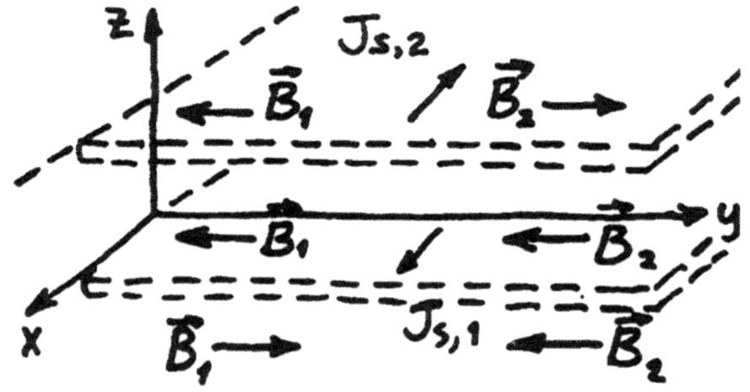

Problem 39 Solution.

Problem

41. A hollow conducting pipe of inner radius a and outer radius b carries a current I parallel to its axis and distributed uniformly through the pipe material (Fig. 30-61). Find expressions for the magnetic field for (a) $r < a$, (b) $a < r < b$, and (c) $r > b$, where r is the radial distance from the pipe axis.

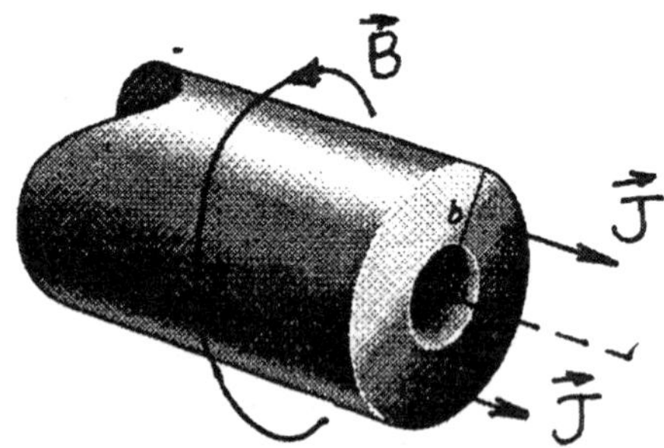

FIGURE 30-61 Problem 41.

Solution

The symmetry argument in the text, for the field of a straight wire, shows that the magnetic field lines (from a very long pipe) are concentric circles, counterclockwise for current out of the page. Ampere's law, for loops along the field lines, gives $2\pi r B = \mu_0 I_{\text{encircled}}$. For uniform current density, $I_{\text{encircled}}$ is proportional to the cross-sectional area of conducting material. Therefore,

$$B(r) = \frac{\mu_0}{2\pi r}\begin{cases} 0, & r < a \\ I\dfrac{(r^2 - a^2)}{(b^2 - a^2)}, & a \leq r \leq b \\ I, & r > b \end{cases}$$

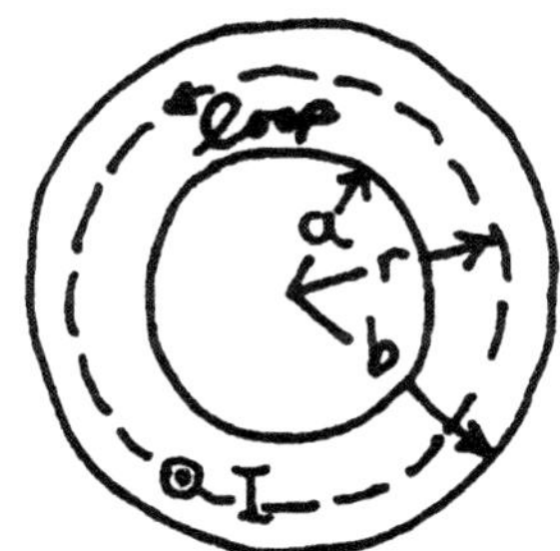

Problem 41 Solution.

Section 30-5: Solenoids and Toroids

Problem

43. A superconducting solenoid has 3300 turns per meter and can carry a maximum current of 4.1 kA. What is the magnetic field strength in the solenoid?

Solution

Equation 30-11 for a long thin solenoid (if applicable) gives $B = \mu_0 nI = (4\pi\times10^{-7}\ \text{N/A}^2)(3300/\text{m})\times(4.1\ \text{kA}) = 17.0\ \text{T}$.

Problem

45. You have 10 m of 0.50-mm-diameter copper wire and a battery capable of passing 15 A through the wire. What magnetic field strengths could you obtain (a) inside a 2.0-cm-diameter solenoid wound with the wire as closely spaced as possible and (b) at the center of a single circular loop made from the wire?

Solution

(a) The length of a solenoid, with one layer of N turns of closely spaced wire of diameter d, is $\ell = Nd$, so the number of turns per unit length is $n = N/\ell = d^{-1} = (0.5\ \text{mm})^{-1} = 2000\ \text{m}^{-1}$. (Although not needed, the value of $N = 10\ \text{m}/2\pi\ \text{cm} = 159$ turns can be used to check that $\ell = 159\times0.5\ \text{mm} = 7.96\ \text{cm}$ is barely long enough, compared to 2 cm, to justify the use of Equation 30-11 as an approximation to B at the solenoid's center.) Then $B = \mu_0 nI = (4\pi\times10^{-7}\ \text{N/A}^2)\times(2000\ \text{m}^{-1})(15\ \text{A}) = 3.77\times10^{-2}\ \text{T}$. (b) The magnetic field at the center of a flat, circular current loop can be found from Equation 30-3, $B = \mu_0 I/2a$. Here, $2\pi a = 10$ m, so $B = \mu_0\pi I/10\ \text{m} = (4\pi^2\times10^{-7}\ \text{N/A}^2)\times(15\ \text{A})/10\ \text{m} = 5.92\times10^{-6}\ \text{T}$.

Problem

47. A toroidal fusion reactor requires a magnetic field that varies by no more than 10% from its central value of 1.5 T. If the minor radius of the toroidal coil producing this field is 30 cm, what is the minimum value for the major radius of the device?

Solution

The central value of the toroidal field is given by Equation 30-12 with $r = R_{\text{maj}}$. At other values of r inside the toroid, the percent difference in field strength is $100(B - B_{\text{maj}})/B_{\text{maj}} = 100(R_{\text{maj}} - r)/r$. The extremes of r are $R_{\text{maj}} \pm R_{\text{min}}$, so it is required that $10 \geq 100\,|[R_{\text{maj}} - (R_{\text{maj}} \pm R_{\text{min}})] \div (R_{\text{maj}} \pm R_{\text{min}})|$, or $R_{\text{maj}} \geq (10 \mp 1)R_{\text{min}}$. The minimum major radius (corresponding to the limit using the smallest value of r, or the lower sign above) is $R_{\text{maj}} \geq 11R_{\text{min}} = 11\times30\ \text{cm} = 3.30\ \text{m}$. (Since 10% is not infinitesimal, differentiation w.r.t. r gives an alternative approximate limit: $|dB/B| = |-dr/r| = R_{\text{min}}/R_{\text{maj}} \leq 10\%$, or $R_{\text{maj}} \geq 3$ m.)

Problem

49. We noted that there is a nonzero magnetic field component outside a solenoid, encircling the device, associated with the component of current flow parallel to the solenoid axis. For a long solenoid of radius R, find an expression for the ratio of this external encircling field just outside the solenoid to the field inside, and show explicitly that this ratio tends to zero as the number of turns per unit length becomes large.

Solution

If the current I to the solenoid enters at one end and leaves at the other, this constitutes a net flow along the solenoid axis. Treating this axial current as a long straight wire, one finds that a magnetic field of $B' = \mu_0 I/2\pi R$, encircling the solenoid just outside its surface, is generated. The field inside the solenoid is $B = \mu_0 nI$, so the ratio is $B'/B = 1/2\pi nR$, which becomes small when the number of turns per unit length, n, is large. (Note: There must be another field outside the solenoid due to its finite length and the fact that the magnetic field lines passing inside are parts of closed loops.)

Section 30-6: Magnetic Matter

Problem

51. When a sample of a certain substance is placed in a 250.0-mT magnetic field, the field inside the sample is 249.6 mT. Find the magnetic susceptibility of the substance. Is it ferromagnetic, paramagnetic, or diamagnetic?

Solution

Equation 30-13 gives the relation between the internal and applied magnetic fields in terms of the relative permeability or the magnetic susceptibility. For the sample described in this problem, the latter is $\chi_M = (B_{\text{int}} - B_{\text{app}})/B_{\text{app}} = (249.6 - 250.0)/250.0 = -1.6\times10^{-3}$. Since $\chi_M < 0$ (or $B_{\text{int}} < B_{\text{app}}$) the material is diamagnetic.

Problem

53. A ferromagnetic material is placed in a 2.5-G magnetic field and the field within the material is determined to be 1.8 T. What is the magnetic susceptibility of this material?

Solution

From Equation 30-13, $\chi_M = (B_{\text{int}}/B_{\text{app}}) - 1 = (1.8\text{ T}/2.5\text{ G}) - 1 \simeq 7.20\times10^3$. (In ferromagnetic materials, $\chi_M \approx \kappa_M$ and both are functions of the applied field and the past history of the sample.)

Paired Problems

Problem

55. Two concentric, coplanar circular current loops have radii a and $2a$. If the magnetic field is zero at their common center, how does the current in the outer loop compare with that in the inner loop?

Solution

The magnetic field strength at the center of a circular current loop is $\mu_0 I/2R$ (Equation 30-3). In order for the net field to cancel, the currents must be in opposite directions and have magnitudes such that $\mu_0 I_{\text{outer}}/2(2a) = \mu_0 I_{\text{inner}}/2a$. Then $I_{\text{outer}} = 2I_{\text{inner}}$.

Problem

57. Figure 30-64 shows a wire of length ℓ carrying current fed by other wires that are not shown. Point A lies on the perpendicular bisector, a distance y from the wire. Adapt the calculation of Example 30-2 to show that the magnetic field at A due to the straight wire alonc has magnitude $\frac{\mu_0 I\ell}{2\pi y\sqrt{\ell^2+4y^2}}$. What is the field direction?

Solution

To find the magnetic field at point A, one may use the same argument as in Example 30-2, except that for a wire of finite length, one integrates from $x = -\ell/2$ to $x = \ell/2$. For current flowing in the positive x direction, **B** is out of the page at point A, with strength

$$B(A) = \int_{-\ell/2}^{\ell/2} \frac{\mu_0 I}{4\pi}\frac{y\,dx}{(x^2+y^2)^{3/2}} = \frac{\mu_0 I y}{4\pi}\left|\frac{x}{y^2\sqrt{x^2+y^2}}\right|_{-\ell/2}^{\ell/2} = \frac{\mu_0 I}{2\pi y}\frac{\ell}{\sqrt{\ell^2+4y^2}}.$$

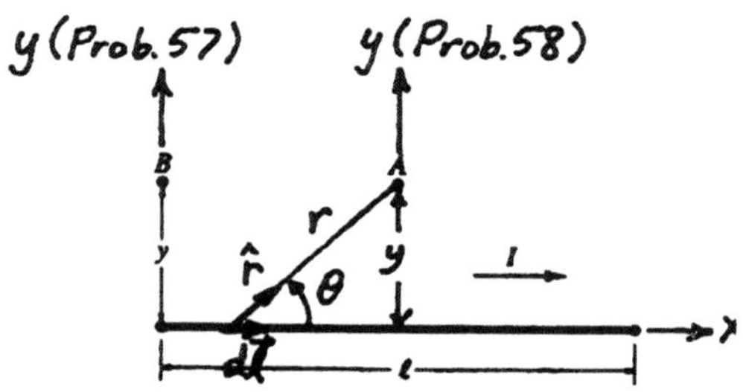

FIGURE 30-64 Problem 57 Solution.

Problem

59. The largest lightning strikes have peak currents around 250 kA, flowing in essentially cylindrical channels of ionized air. How far from such a flash would the resulting magnetic field be equal to Earth's magnetic field strength, about 50 μT?

Solution

Supposing that the cylindrical channel of ionized air acts like a long straight wire, we can use Equation 30-5 to estimate the distance: $y = \mu_0 I/2\pi B = (2\times10^{-7}\text{ N/A}^2)(250\text{ kA})/(50\ \mu\text{T}) = 1$ km.

Problem

61. A coaxial cable like that shown in Fig. 30-60 consists of a 1.0-mm-diameter inner conductor and an outer conductor of inner diameter 1.0 cm and 0.20 mm thickness. A 100-mA current flows down the center conductor and back along the outer conductor. Find the magnetic field strength (a) 0.10 mm, (b) 5.0 mm, and (c) 2.0 cm from the cable axis.

Solution

For a long, straight cable, the magnetic field can be found from Ampere's law. The field lines are cylindrically symmetric and form closed loops, hence must be concentric circles, which we also choose as amperian loops. Take positive circulation counterclockwise so that positive current is out of the page. Then $\oint_c \mathbf{B}\cdot d\boldsymbol{\ell} = 2\pi rB = \mu_0 I_{\text{encircled}}$. Assume that the current density in each conductor is uniform; i.e., the current is proportional to the cross-sectional area. We may calculate $I_{\text{encircled}}$ in four regions of space. (a) For $r \le R_a, I_{\text{encircled}} = I(\pi r^2/\pi R_a^2) = Ir^2/R_a^2$, so $B = \mu_0 Ir/2\pi R_a^2$. (b) For $R_a \le r \le R_b$, $I_{\text{encircled}} = I$, so $B = \mu_0 I/2\pi r$. (Although not asked for, for $R_b \le r \le R_c$,

$$I_{\text{encircled}} = I - I\frac{\pi(r^2 - R_b^2)}{\pi(R_c^2 - R_b^2)} = I\left(\frac{R_c^2 - r^2}{R_c^2 - R_b^2}\right), \text{ so}$$

$$B = \frac{\mu_0 I}{2\pi(R_c^2 - R_b^2)}\left(\frac{R_c^2}{r} - r\right).$$

The outer radius R_c is the inner radius plus the thickness of the outer conductor.) (c) Finally, for $r \ge R_c, I_{\text{encircled}} = 0$, so $B = 0$. Numerically, $R_a = 0.5$ mm, $R_b = 5$ mm, $R_c = 5.2$ mm, and $I = 0.1$ A, so inside the inner conductor (a) when $r = 0.1 \text{ mm} \le R_a, B = (2\times10^{-7}\ \text{T·m/A})\times (0.1\ \text{A})(0.1\ \text{mm})/(0.5\ \text{mm})^2 = 8\ \mu\text{T}$, between the two conductors (b) when $R_a \le r = 5 \text{ mm} \le R_b, B = (2\times10^{-7}\ \text{T·m/A})(0.1\ \text{A})/(5\ \text{mm}) = 4\ \mu\text{T}$; and outside the outer conductor (c) when $r = 2 \text{ cm} \ge R_c, B = 0$.

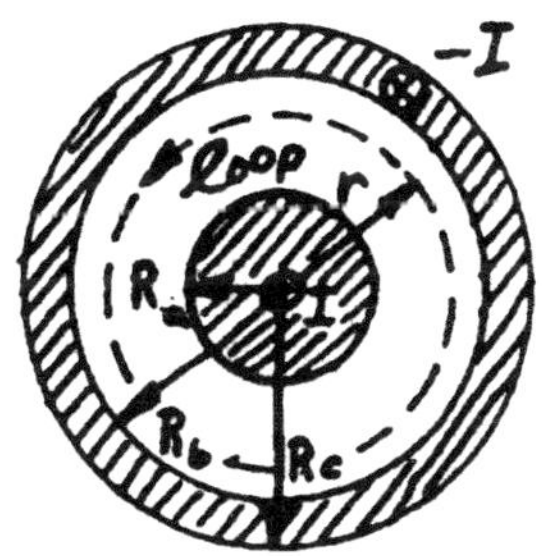

Problem 61 Solution.

Supplementary Problems

Problem

63. A circular wire loop of radius 15 cm and negligible thickness carries a 2.0-A current. Use suitable approximations to find the magnetic field of this loop (a) in the loop plane, 1.0 mm outside the loop, and (b) on the loop axis, 3.0 m from the loop center.

Solution

(a) As mentioned in the paragraph following Equation 30-4, the calculation of the magnetic field from a circular current loop, at points not on its axis, is difficult. Fortunately, at a distance of 1 mm from a loop of radius 15 cm, the field is approximately that of a long, straight wire, Equation 30-5 gives $B \approx \mu_0 I/2\pi y = (2\times10^{-7}\ \text{N/A}^2)(2\ \text{A})/(10^{-3}\ \text{m}) = 4$ G. (b) Since 3 m $\gg$ 15 cm, the approximation $x \gg a$ in Equation 30-4 is justified. Then $B \approx \mu_0 Ia^2/2|x|^3 = (2\pi\times10^{-7}\ \text{N/A}^2)(2\ \text{A})(0.15\ \text{m})^2/(3\ \text{m})^3 = 1.05\times10^{-5}$ G.

Problem

65. A long, hollow conducting pipe of radius R and length ℓ carries a uniform current I flowing around the pipe, as shown in Fig. 30-66. Find expressions for the magnetic field (a) inside and (b) outside the pipe. *Hint:* What configuration does this pipe resemble?

Solution

The current distribution is similar to a solenoid, where the number of turns per unit length and the current in each turn are related to the total current in the pipe by $n\ell I_t = I$. Therefore (see Section 30-5), the field is approximately that of an infinite solenoid: $B = \mu_0 nI_t = \mu_0 I/\ell$ inside, and $B = 0$ outside, directed parallel to the axis to the left in Fig. 30-66.

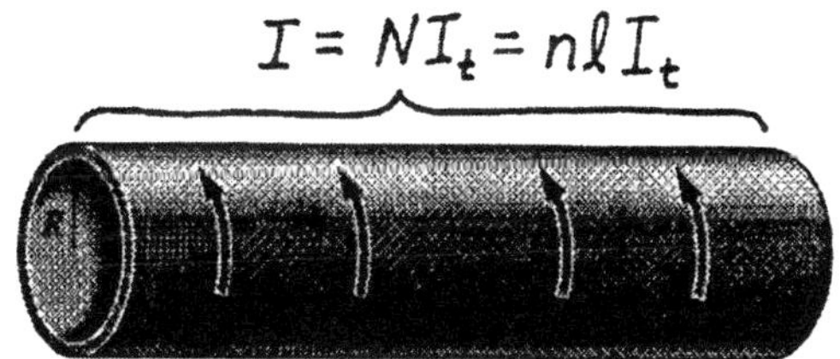

FIGURE 30-66 Problem 65 Solution.

Problem

67. A wide, flat conducting spring of spring constant $k = 20$ N/m and negligible mass consists of two 6.0-cm-diameter turns, as shown in Fig. 30-67. In

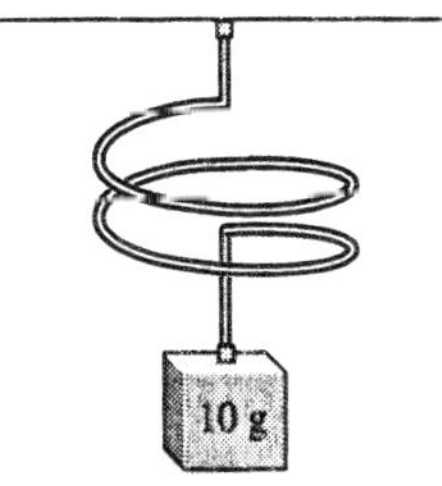

FIGURE 30-67 Problem 67.

its unstretched configuration the coils are nearly touching. A 10-g mass is hung from the spring, and at the same time a current I is passed through it. The spring stretches 2.0 mm. Find I, assuming the coils remain close enough to be treated as parallel wires.

Solution

The length of one turn, $\ell = 6\pi$ cm, is large compared to the separation of the two turns, $d = 2$ mm, so that Equation 30-6 can be used to find the magnetic force in the spring. $F_{\text{mag}} = \mu_0 I^2 \ell / 2\pi d = (2\times10^{-7}\text{ N/A}^2)\times I^2(6\pi\text{ cm})/2\text{ mm} = 1.88 I^2\times10^{-5}\text{ N/A}^2$. The elastic force in the spring, in the same direction as F_{mag}, is $F_{\text{el}} = kd = (20\text{ N/m})(2\text{ mm}) = 4\times10^{-2}$ N. At equilibrium, $F_{\text{el}} + F_{\text{mag}} = mg = (10^{-2}\text{ kg})\times(9.8\text{ m/s}^2) = 9.8\times10^{-2}$ N. Thus, $1.88 I^2\times 10^{-5}\text{ N/A}^2 = 5.8\times10^{-2}$ N, or $I = 55.5$ A.

Problem

69. A disk of radius a carries a uniform surface charge density σ, and is rotating with angular speed ω about the central axis perpendicular to the disk. Show that the magnetic field at the disk's center is $\frac{1}{2}\mu_0\sigma\omega a$.

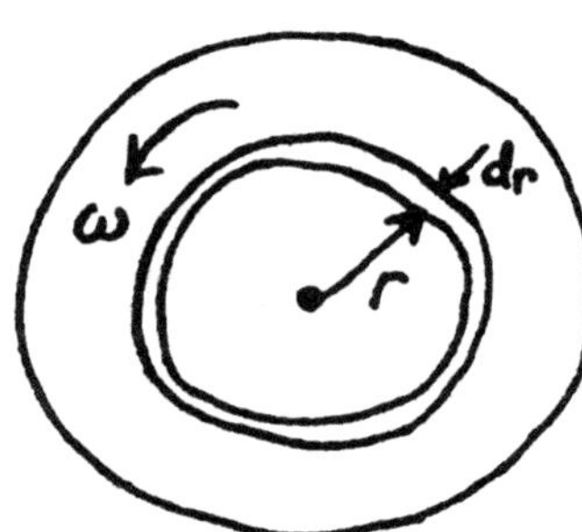

Problem 69 Solution.

Solution

The disk may be considered to be composed of rings of radius r, thickness dr, and charge $dq = 2\pi r\, dr\cdot\sigma$. Each ring represents a circular current loop (see hint in problem 29-73) $dI = dq/T = dq/(2\pi/\omega) = \omega\sigma r\, dr$, which produces a magnetic field strength $dB = \mu_0\, dI/2r = \frac{1}{2}\mu_0\omega\sigma\, dr$ at the center of the disk, directed out of the page, as sketched for positive charge density. The total field strength is $B = \int_0^a dB = \frac{1}{2}\mu_0\omega\sigma\int_0^a dr = \frac{1}{2}\mu_0\omega\sigma a$.

Problem

71. Work Example 30-2 by expressing all variables in terms of the angle θ and integrating over the appropriate range in θ.

Solution

In Example 30-2 and Fig. 30-9, $\cos\theta = -x/r$, $\tan\theta = -y/x$, so $d(\tan\theta) = d\theta/\cos^2\theta = y\, dx/x^2$. Then $dx = (x^2/\cos^2\theta)\, d\theta/y = r^2\, d\theta/y$. Thus, we can write the field element (out of the page) as

$$dB = \frac{\mu_0 I}{4\pi}\frac{dx\sin\theta}{r^2} = \frac{\mu_0 I}{4\pi}\left(\frac{r^2 d\theta}{y}\right)\frac{\sin\theta}{r^2} = \frac{\mu_0 I}{4\pi y}\sin\theta\, d\theta.$$

The limits of integration $x = -\infty$ to $+\infty$ correspond to $\theta = 0$ to π (or 180°), hence

$$B = \frac{\mu_0 I}{4\pi y}\int_0^\pi \sin\theta\, d\theta = \frac{\mu_0 I}{4\pi y}\bigg| -\cos\theta\bigg|_0^\pi = \frac{\mu_0 I}{2\pi y},$$

which is the same as Equation 30-5.

CHAPTER 31 ELECTROMAGNETIC INDUCTION

ActivPhysics can help with these problems: Activities 13.9, 13.10

Sections 31-2 and 31-3: Faraday's Law and Induction and the Conservation of Energy

Problem

1. Show that the volt is the correct SI unit for the rate of change of magnetic flux, making Faraday's law dimensionally correct.

Solution

The units of $d\phi_B/dt$ are $\text{T·m}^2/\text{s} = (\text{N/A·m})(\text{m}^2/\text{s}) = (\text{N·m/A·s}) = \text{J/C} = \text{V}$.

Problem

3. Find the magnetic flux through a circular loop 5.0 cm in diameter oriented with the loop normal at 30° to a uniform 80-mT magnetic field.

Solution

For a stationary plane loop in a uniform magnetic field, the integral for the flux in Equation 31-1 is just $\phi_B = \mathbf{B} \cdot \mathbf{A} = BA\cos\theta = (80 \text{ mT})\pi(2.5 \text{ cm})^2 \cos 30° = 1.36 \times 10^{-4}$ Wb. (The SI unit of flux, T·m^2, is also called a weber, Wb.)

Problem

5. A conducting loop of area 240 cm^2 and resistance 12 Ω lies at right angles to a spatially uniform magnetic field. The loop carries an induced current of 320 mA. At what rate is the magnetic field changing?

Solution

The flux through a stationary loop perpendicular to a magnetic field is $\phi_B = BA$ (see Problem 3), so Faraday's law (Equation 31-2) and Ohm's law (Equation 27-6) relate this to the magnitude of the induced current: $I = |\mathcal{E}/R| = |d\phi_B/dt|/R = A\,|dB/dt|/R$. Therefore $|dB/dt| = IR/A = (320 \text{ mA})(12\ \Omega) \div (240 \text{ cm}^2) = 160$ T/s.

Problem

7. A conducting loop with area 0.15 m^2 and resistance 6.0 Ω lies in the x-y plane. A spatially uniform magnetic field points in the z direction. The field varies with time according to $B_z = at^2 - b$, where $a = 2.0 \text{ T/s}^2$ and $b = 8.0$ T. Find the loop current (a) when $t = 3.0$ s and (b) when $B_z = 0$.

Solution

The reasoning in the solution to Problem 5 shows that $|I| = A\,|dB/dt|/R = A\,(2at)/R = (0.15 \text{ m}^2) \times (2 \times 2 \text{ T/s}^2)t/(6\ \Omega) = 10^{-1}t$ (A/s). (a) For $t = 3$ s, $|I| = 0.3$ A, and (b) for $t = \sqrt{b/a} = \sqrt{8 \text{ T}/(2 \text{ T/s}^2)} = 2$ s, $|I| = 0.2$ A. (Note: In this problem, there is enough information to also specify the direction of I. Choose the x-y axes as shown and the z-axis out of the page. For positive normal to the loop along the z-axis, a positive sense of circulation for the induced current is CCW. Then $I = \mathcal{E}/R = -(1/R)d(\mathbf{B} \cdot \mathbf{A})/dt = -(A/R)(dB_z/dt) = -(A/R)(2at)$. Negative currents are CW around the z-axis.)

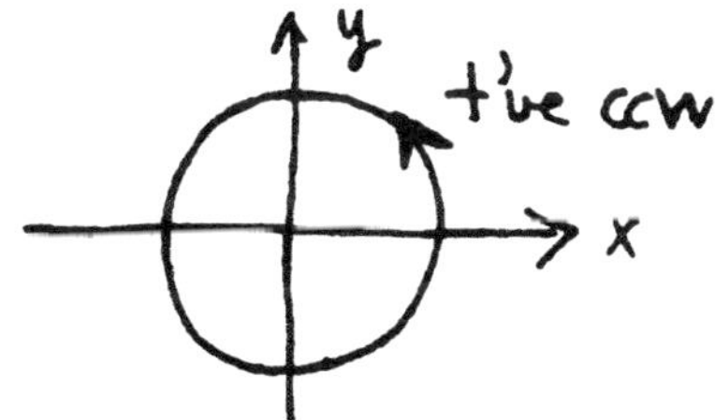

Problem 7 Solution.

Problem

9. A square wire loop of side ℓ and resistance R is pulled with constant speed v from a region of no magnetic field until it is fully inside a region of constant, uniform magnetic field $\mathbf{B}$ perpendicular to the loop plane. The boundary of the field region is parallel to one side of the loop. Find an expression for the total work done by the agent pulling the loop.

Solution

The loop can be treated analogously to the situation analyzed in Section 31-3, under the heading "Motional EMF and Lenz's Law"; instead of exiting, the loop is entering the field region at constant velocity. All quantities have the same magnitudes, except the current in the loop is CCW instead of CW, as in Fig. 31-13. Since the applied force acts over a displacement equal to the side-length of the loop, the work done can be calculated directly: $W_{app} = \mathbf{F}_{app} \cdot \boldsymbol{\ell} = (I\ell B)\ell = I\ell^2 B$. But, $I = \mathcal{E}/R = |d\phi_B/dt|/R = d/dt(B\ell x)/R = B\ell v/R$, as before, so $W_{app} = B^2\ell^3 v/R$. [Alternatively, the work can be calculated from the conservation of energy: $I = B\ell v/R$, $\mathcal{P}_{diss} = I^2 R = (B\ell v)^2/R$, and $W_{app} = \mathcal{P}_{diss}t = [(B\ell v)^2/R](\ell/v)$.]

Problem

11. In Fig. 31-26 the loop radius is 15 cm, and the magnetic field is decreasing at the rate of 550 T/s. If the gap width is small compared with the loop circumference, what is the voltage across the gap?

Solution

If the gap is small, essentially the entire induced emf around the loop appears across it (as explained after Example 31-7 in the sub section of the text on "Induction in Open Circuits"), so $|\mathcal{E}| = |d\phi_B/dt| = \pi R^2 |dB/dt| = \pi(0.15 \text{ m})^2(550 \text{ T/s}) = 38.9$ V. (The polarity is as mentioned in the text.)

Problem

13. The wingspan of a 747 jetliner is 60 m. If the plane is flying at 960 km/h in a region where the vertical component of Earth's magnetic field is 0.20 G, what emf develops between the plane's wingtips?

Solution

A motional emf causes electrons in the wing to drift until equilibrium with the electrostatic field from accumulated wingtip charges is achieved. The magnetic and electric forces on an electron have magnitudes $F_{mag} = |-e\mathbf{v} \times \mathbf{B}| = evB_\perp$ (we suppose that the 747 is flying horizontally so only the vertical magnetic field gives a force parallel to the wingspan), and $F_{el} = |-e\mathbf{E}_s| = e\text{V}/\ell$, where V is the potential difference between the wingtips. At equilibrium, $evB = e\text{V}/\ell$, or $\text{V} = B\ell v = (2\times10^{-5} \text{ T})(60 \text{ m})\times(960 \text{ m}/3.6 \text{ s}) = 320$ mV. (Motional emf's like this need to be considered in rocket measurements of ionospheric electric fields.)

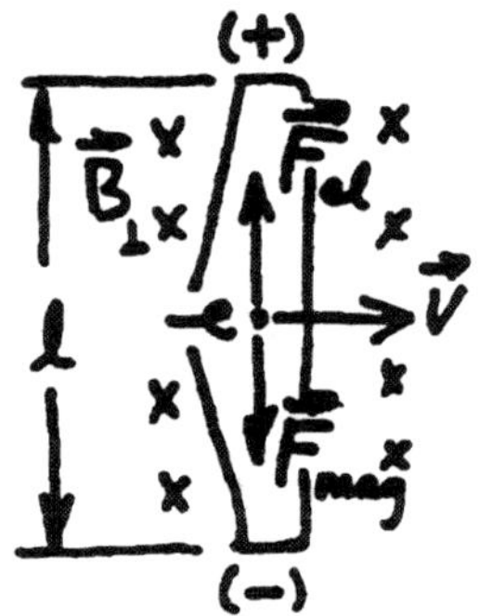

Problem 13 Solution.

Problem

15. In Example 31-2 take $a = 1.0$ cm, $w = 3.5$ cm, and $\ell = 6.0$ cm. Suppose the rectangular loop is a conductor with resistance 50 mΩ and that the current I in the long wire is increasing at the rate of 25 A/s. Find the induced current in the loop. In what direction does it flow?

Solution

The normal to the loop in Fig. 31-6 was taken to be in the direction of the magnetic field of the wire, or into the page, so the positive sense of circulation around the loop is clockwise (from the right-hand rule). Faraday's and Ohm's laws, together with the result of Example 31-2, give an induced current the loop of $\mathcal{E}/R = (-d\phi_B/dt)/R = -(\mu_0\ell/2\pi R)\ln[(a + w)/a]\times(dI/dt) = -(2\times10^{-7} \text{ N/A}^2)(6 \text{ cm})(25 \text{ A/s})\times\ln[4.5/1]/(50 \text{ m}\Omega) = -9.02$ μA. A negative current is counterclockwise in the loop.

Problem

17. A square conducting loop of side $s = 0.50$ m and resistance $R = 5.0$ Ω moves to the right with speed $v = 0.25$ m/s. At time $t = 0$ its rightmost edge enters a uniform magnetic field B=1.0 T pointing into the page, as shown in Fig. 31-46. The magnetic field covers a region of width $\omega = 0.75$ m. Plot (a) the current and (b) the power dissipation in the loop as functions of time, taking a clockwise current as positive and covering the time until the entire loop has exited the field region.

Solution

Let x be the distance between the right side of the loop and the left edge of the field region. Take $t = 0$ when $x = 0$, so that $x = vt$. The loop enters the field region at $t = 0$, is completely within the region for t between $\ell/v = 2$ s and $w/v = 3$ s, and is out of the region for $t \geq (w + \ell)/v = 5$ s.
The area of loop overlapping the field region increases linearly from 0 to ℓ^2, stays constant at ℓ^2, then decreases to 0 between these times. (We use ℓ for side

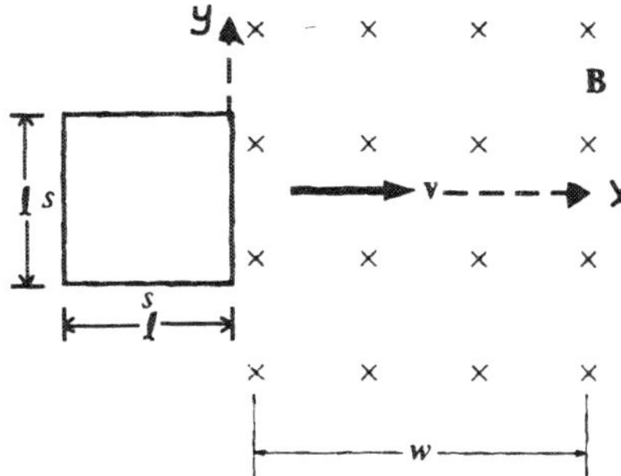

FIGURE 31-46 Problem 17 Solution.

length to avoid confusion with time units.). Thus,

$$\phi_B = BA = B\ell^2 \begin{cases} 0 \\ vt/\ell \\ 1 \\ (w+\ell-vt)/\ell \\ 0 \end{cases}$$

$$= 0.25\text{ Wb} \begin{cases} 0, & t \leq 0 \\ 0.5t, & 0 \leq t \leq 2 \\ 1, & 2 \leq t \leq 3 \\ 0.5(5-t), & 3 \leq t \leq 5 \\ 0, & 5 \leq t \end{cases}$$

(We substituted the given numerical values and used SI units for flux, with time t in seconds, see solution to Problem 3.)

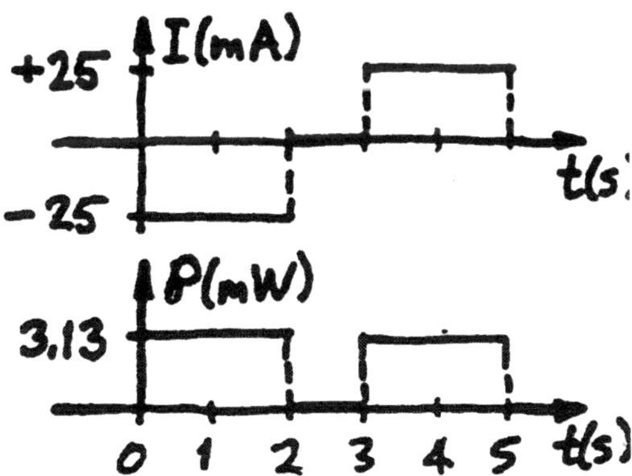

Problem 17 Solution.

(a) The induced current (positive clockwise) is given by Faraday's and Ohm's laws:

$$I = -\frac{1}{R}\frac{d\phi_B}{dt} = 25\text{ mA} \begin{cases} 0, & t \leq 0 \\ -1, & 0 \leq t \leq 2 \\ 0, & 2 \leq t \leq 3 \\ +1, & 3 \leq t \leq 5 \\ 0, & 5 \leq t \end{cases}$$

(b) The power dissipated, I^2R, is $(\pm 25\text{ mA})^2(5\ \Omega) =$ 3.13 mW when the current is not zero.

Problem

19. A solenoid 2.0 m long and 30 cm in diameter consists of 5000 turns of wire. A 5-turn coil with negligible resistance is wrapped around the solenoid and connected to a 180-Ω resistor, as shown in Fig. 31-47. The direction of the current in the solenoid is such that the solenoid's magnetic field points to the right. At time $t = 0$ the solenoid current begins to decay exponentially, being given by $I = I_0e^{-t/\tau}$, where $I_0 = 85$ A, $\tau = 2.5$ s, and t is the time in seconds. (a) What is the direction of the current in the resistor as the solenoid current decays? What is the value of the resistor current at (b) $t = 1.0$ s and (c) t=5.0 s?

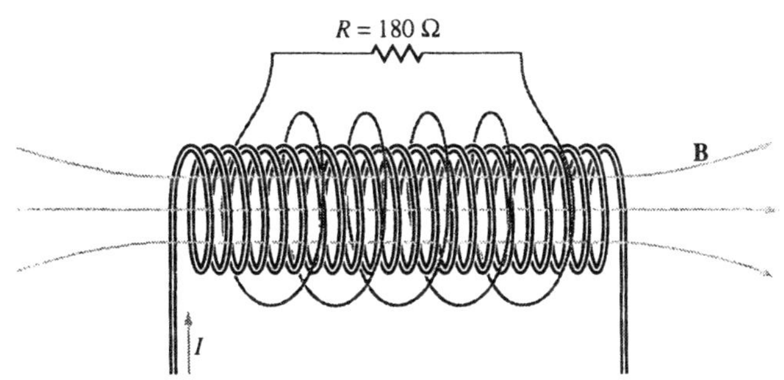

FIGURE 31-47 Problem 19.

Solution

(a) From Lenz's law, the direction of the induced current in the coil, I_c, must oppose the decrease in the solenoid's field, B_s, to the right, so the induced field due to I_c must also be to the right. Thus, the right-hand rule for the field of the coil requires that I_c flow from left to right in the 180 Ω resistor in Fig. 31-47. (b) The field inside the solenoid ($B_s = \mu_0(N_s/\ell)I_s$) links each turn of the coil, so the total flux through the latter is $\phi_B = N_cB_sA_s$. (Recall that approximately none of the solenoid's field lines go through the coil outside the solenoid's cross-sectional area.) From Faraday's and Ohm's laws, the induced current in the coil has magnitude $I_c = |\mathcal{E}|/R =$ $(1/R)\,|-d\phi_B/dt| = (\mu_0N_cN_s/\ell)\frac{1}{4}\pi D_s^2(1/R)\,|-dI_s/dt|$. With $I_s = I_0e^{-t/\tau}$, $|dI_s/dt| = \left|-(I_0/\tau)e^{-t/\tau}\right|$, and the given numerical data,

$$I_c = \left(\pi^2\times10^{-7}\frac{\text{N}}{\text{A}^2}\right)\left(\frac{5000\times5}{2\text{ m}}\right)\frac{(0.3\text{ m})^2}{(180\ \Omega)}\left(\frac{85\text{ A}}{2.5\text{ s}}\right)e^{-t/2.5\text{ s}}$$
$$= (210\ \mu A)e^{-t/2.5\text{ s}}.$$

At $t = 1$ s, $I_c = (210\ \mu\text{A})e^{-0.4} = 141\ \mu\text{A}$, and (c) at $t = 5$ s, $I_c = (210\ \mu\text{A})e^{-2} = 28.4\ \mu\text{A}$.

Problem

21. (a) Find an expression for the resistor current in Problem 19 if the solenoid current is given by $I = I_0 \sin \omega t$, where $I_0 = 85$ A and $\omega = 210$ s^{-1}. (b) What is the peak current in the resistor? (c) What is the resistor current when the solenoid current is a maximum?

Solution

(a) In Problem 19, $I_c = \mu_0(N_cN_s/\ell)\frac{1}{4}\pi(D_s^2/R)\times(-dI_s/dt)$ was the expression for the current in the coil, taken as positive from left to right in the resistor. When the current in the solenoid is $I_s = I_0 \sin \omega t$, $dI_s/dt = \omega I_0 \cos \omega t$, so $I_c = -(\mu_0\pi/4)(N_cN_s/\ell)\times(D_s^2/R)\omega I_0 \cos \omega t \equiv -I_{\text{peak}} \cos \omega t$. (b) Numerically, $I_{\text{peak}} = (\pi^2\times10^{-7}\text{N/A}^2)(5\times5000/2\text{ m})(0.3\text{ m})^2\times(210\text{ s}^{-1})(85\text{ A})/(180\ \Omega) = 110$ mA. (c) When $\sin \omega t$ is a maximum, $\cos \omega t$ is zero.

Problem

23. In the preceding problem, what is the first time after $t = 0$ when the loop current will be zero?

Solution

Take the normal to the loop along the z-axis, so that the positive sense of circulation for the induced emf and current is CCW (looking down on the x-y plane) as shown. The flux through the loop is $\phi_B = \mathbf{B}\cdot\mathbf{A} = (B_0 \sin \omega t\ \hat{\mathbf{k}})(A\hat{\mathbf{k}}) = B_0 A \sin \omega t$, so the induced current is $I = \mathcal{E}/R = (-d\phi_B/dt)/R = -(\omega B_0 A/R)\times \cos \omega t$. The first time cos ωt is zero, after $t = 0$ is when $\omega t = \pi/2$, or $t = \pi/2\omega = \pi/20$ s^{-1} = 157 ms.

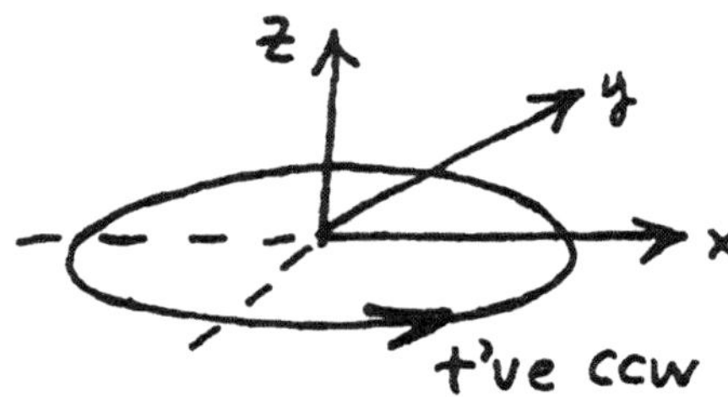

Problem 23 Solution.

Problem

25. A credit-card reader extracts information from the card's magnetic stripe as it is pulled past the reader's head. At some instant the card motion results in a magnetic field at the head that is changing at the rate of 450 μT/ms. If this field passes perpendicularly through a 5000-turn head coil 2.0 mm in diameter, what will be the induced emf?

Solution

The magnetic flux through the coil in the reader's head is changing at a rate of $NA\ dB/dt = (5000)\pi(1\text{ mm})^2(450\ \mu\text{T/ms}) = 7.07$ mV. According to Faraday's law, this is equal to the magnitude of the induced emf.

Problem

27. Figure 31-49 shows a pair of parallel conducting rails a distance ℓ apart in a uniform magnetic field $\mathbf{B}$. A resistance R is connected across the rails, and a conducting bar of negligible resistance is being pulled along the rails with velocity $\mathbf{v}$ to the right. (a) What is the direction of the current in the resistor? (b) At what rate must work be done by the agent pulling the bar?

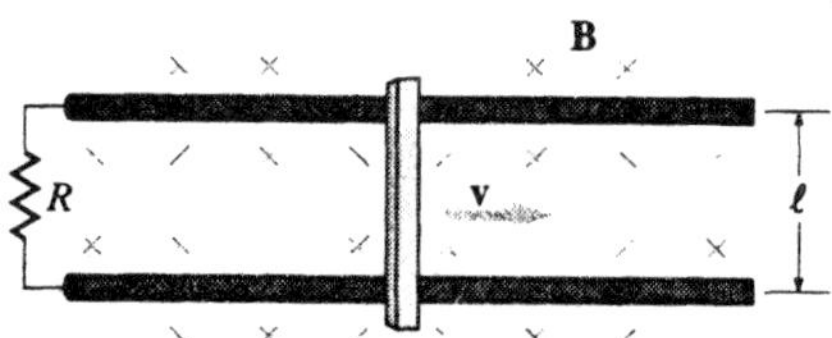

FIGURE 31-49 Problem 27.

Solution

(a) The force on a (hypothetical) positive charge carrier in the bar, $q\mathbf{v}\times\mathbf{B}$, is upward in Fig. 31-49, so current will circulate CCW around the loop containing the bar, the resistor, and the rails (i.e., downward in the resistor). (The force per unit positive charge is the motional emf in the bar.) Alternatively, since the area enclosed by the circuit, and the magnetic flux through it, are increasing, Lenz's law requires that the induced current oppose this with an upward induced magnetic field. Thus, from the right-hand rule, the induced current must circulate CCW. (Take the positive sense of circulation around the circuit CW, so that the normal to the area is in the direction of $\mathbf{B}$, into the page.) (b) In Example 31-4, which analyzed the same situation, the current in the bar was found to be $I = |\mathcal{E}|/R = B\ell v/R$. Since this is perpendicular to the magnetic field, the magnetic force on the bar is $F_{\text{mag}} = I\ell B$ (to the left in Fig. 31-49). The agent pulling the bar at constant velocity must exert an equal force in the direction of $\mathbf{v}$, and therefore does work at the rate $\mathbf{F}\cdot\mathbf{v} = I\ell Bv = (B\ell v)^2/R$. (Note: The conservation of energy requires that this equal the rate energy is dissipated in the resistor (we neglected the resistance of the bar and the rails), $I^2R = (B\ell v/R)^2R$.)

Problem

29. A battery of emf $\mathcal{E}$ is inserted in series with the resistor in Fig. 31-49, with its positive terminal toward the top rail. The bar is initially at rest, and now no agent pulls it. (a) Describe the bar's subsequent motion. (b) The bar eventually reaches a constant speed. Why? (c) What is that constant speed? Express in terms of the magnetic field, the battery emf, and the rail spacing ℓ. Does the resistance R affect the final speed? If not, what role does it play?

Solution

(a) The battery causes a CW current (downward in the bar) to flow in the circuit composed of the bar, resistor, and rails. (For positive circulation CW, the right-hand rule gives a positive normal to the area bounded by the circuit into the page, so that the flux $\phi_B = \mathbf{B}\cdot\mathbf{A} = B\ell x$ is positive. The length of the circuit is x, as in Example 31-4.) Thus, there is a magnetic force $\mathbf{F}_{\text{mag}} = I\boldsymbol{\ell}\times\mathbf{B} = I\ell B$ to the right, which accelerates the bar in that direction. (Any other forces on the bar are assumed to cancel, or be negligible.) An induced emf opposes the battery ($\mathcal{E}_i = -d\phi_B/dt = -B\ell v$, as in Example 31-4, the negative sign indicating a CCW sense in the circuit) so the instantaneous current is $I(t) = (\mathcal{E}+\mathcal{E}_i)/R = (\mathcal{E} - B\ell v)/R$. Thus, as v increases, I (and the accelerating force) decreases. (b) Eventually ($t\to\infty$), $I(\infty) = 0$, $F_{\text{mag}} = I(\infty)\ell B = 0$, and the velocity v_∞ stays constant. (c) When $I(\infty) = 0$, $\mathcal{E} - B\ell v_\infty = 0$, so $v_\infty = \mathcal{E}/B\ell$. Although v_∞ doesn't depend on the resistance, the value of R does affect how rapidly v approaches v_∞. For large R, I charges slowly and v takes a long time to reach v_∞. (The equation of motion of the bar (mass m) is $m(dv/dt) = I\ell B = (\mathcal{E} - B\ell v)\ell B/R$, which can be separated: $dv/(v_\infty - v) = (\ell^2B^2/mR)dt$. For $v_0 = 0$, this integrates to $\ln(1 - v/v_\infty) = -\ell^2B^2t/mR$, or $v = v_\infty(1 - e^{-\ell^2B^2t/mR})$. The time constant, $\tau = mR/\ell^2B^2$, depends on the resistance.)

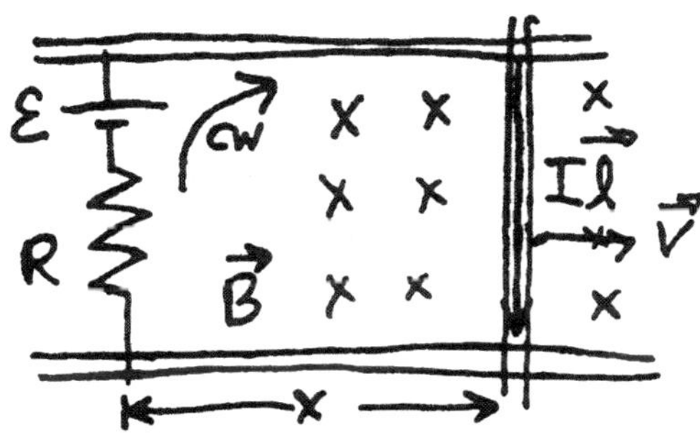

Problem 29 Solution.

Problem

31. A pair of parallel conducting rails 10 cm apart lie at right angles to a uniform magnetic field **B** of magnitude 2.0 T, as shown in Fig. 31-51. A 5.0-Ω and a 10-Ω resistor lie across the rails and are free to slide along them. (a) The 5-Ω resistor is held fixed, and the 10-Ω resistor is pulled to the right at 50 cm/s. What are the direction and magnitude of the induced current? (b) Now the 10-Ω resistor is held fixed, and the 5-Ω resistor is pulled to the left at 50 cm/s. What are the direction and magnitude of the induced current? (c) What is the power dissipation in the 10-Ω resistor in both cases?

Solution

(a) Let x be the distance between the resistors. If we take the normal to the area bounded by the resistors and the rails into the page (positive circulation clockwise), then the flux linking the circuit is $\phi_B = BA = B\ell x$, so Faraday's law gives the magnitude of the induced emf as $\mathcal{E}_i = -d\phi_B/dt = -B\ell(dx/dt) = -B\ell v = -(2\text{ T})(0.1\text{ m})(0.5\text{ m/s}) = -0.1\text{ V}$. The current, therefore, is $I = \mathcal{E}_i/R_{\text{tot}} = -0.1\text{ V} \div (5+10)\ \Omega = -6.67\text{ mA}$. (Lenz's law also gives the direction of I, which opposes the increase in flux, as counterclockwise.) (b) The relative velocity between the loop and the field is the same as in part (a), thus the induced current is the same. (c) $\mathcal{P}_{10\,\Omega} = I^2R_{10\,\Omega} = (6.67\text{ mA})^2(10\ \Omega) = 0.444\text{ mW}$. (We assumed the rails have negligible resistance in calculating I.)

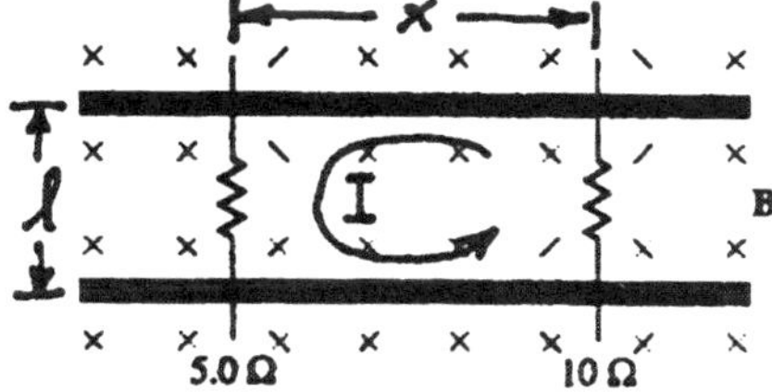

FIGURE 31-51 Problem 31 Solution.

Problem

33. In Fig. 31-49, take $\ell = 10$ cm, $B = 0.50$ T, $R = 4.0\ \Omega$, and $v = 2.0$ m/s. Find (a) the current in the resistor, (b) the magnetic force on the bar, (c) the power dissipation in the resistor, and (d) the mechanical work done by the agent pulling the bar. Compare your answers to (c) and (d).

Solution

The situation is like that described in Example 31-4 and the solution to Problem 27. (a) $I = \mathcal{E}/R = B\ell v/R = (0.5\text{ T})(0.1\text{ m})(2\text{ m/s})/4\ \Omega = 25\text{ mA}$.

(Neglect the resistance of the bar and rails.) (b) $F_{mag} = I\ell B = (25\text{ mA})(0.1\text{ m})(0.5\text{ T}) = 1.25\times10^{-3}\text{ N}$. (c) $\mathcal{P}_J = I^2R = (25\text{ mA})^2(4\ \Omega) = 2.5\text{ mW}$. (d) The agent pulling the bar must exert a force equal in magnitude to F_{mag} and parallel to **v**. Therefore, it does work at a rate $Fv = (1.25\times10^{-3}\text{ N})(2\text{ m/s}) = 2.5\text{ mW}$. The conservation of energy requires the answers to parts (c) and (d) to be equal.

Problem

35. A circular loop 40 cm in diameter is made from a flexible conductor and lies at right angles to a uniform 12-T magnetic field. At time $t = 0$ the loop starts to expand, its radius increasing at the rate of 5.0 mm/s. Find the induced emf in the loop (a) at $t = 1.0$ s and (b) at $t = 10$ s.

Solution

The flux through the loop is $\phi_B = BA = B\pi r^2$ (if we take the normal to the loop's area in the direction of **B**), so Faraday's law gives $\mathcal{E}_i = -d\phi_B/dt = -2\pi Br(dr/dt)$. For $r(t) = r_0 + (5\text{ mm/s})t$, $\mathcal{E}_i = -2\pi(12\text{ T})\times(5\text{ mm/s})(20\text{ cm} + 5t\text{ mm/s})$. At $t = 1$ s, $\mathcal{E}_i = -77.3$ mV, and (b) at $t = 10$ s, $\mathcal{E}_i = -94.2$ mV. The minus sign means that $\mathcal{E}_i$ opposes the increase in flux; a right-hand screw rotated in the sense of $\mathcal{E}_i$ would advance in a direction opposite to **B**.

Section 31-4: Induced Electric Fields

Problem

37. Find the electric force on a 50-μC charge inside the solenoid of Problem 18, if the charge is 5.0 cm from the solenoid axis.

Solution

The induced electric field inside a long thin solenoid is the subject of the exercise accompanying Example 31-9, where a similar argument gives $E = -\frac{1}{2}r(dB/dt)$. Such a field would produce a force on a point charge q, of magnitude $\frac{1}{2}qr(dB/dt) = \frac{1}{2}\mu_0 nqr(dI/dt)$, since $B = \mu_0 nI$. For the solenoid in Problem 18, this is $\frac{1}{2}(4\pi\times10^{-7}\text{ N/A}^2)(2000/2\text{ m})(50\ \mu\text{C})(5\text{ cm})\times(10^3\text{ A/s}) = 1.57\ \mu$N.

Problem

39. A uniform magnetic field points into the page in Fig. 31-54. In the same region an electric field points straight up, but increases with position at the rate of 10 V/m^2 as you move to the right. Apply Faraday's law to a rectangular loop to show that the magnetic field must be changing with time, and calculate the rate of change.

Solution

Assume that there are no sources for **E** other than the changing **B**. Choose coordinate axes and a rectangular loop as shown superposed on Fig. 31-54. The clockwise direction of circulation is such that $d\mathbf{A}$ is parallel to **B** (into the page). The electric field is $\mathbf{E} = [E_0 + (10\text{ V/m}^2)x]\hat{\mathbf{j}}$. This is perpendicular to the sides of width w, and parallel or antiparallel to the sides of length ℓ, located at x_1 and x_2 respectively. Thus, $\oint \mathbf{E}\cdot d\ell = E_1\ell - E_2\ell = [E_0 + (10\text{ V/m}^2)x_1 - E_0 - (10\text{ V/m}^2)x_2]\ell = -(10\text{ V/m}^2)w\ell$, since $x_2 - x_1 = w$. The flux through the loop is $\int \mathbf{B}\cdot d\mathbf{A} = BA = Bw\ell$, so Faraday's law gives $\oint \mathbf{E}\cdot d\boldsymbol{\ell} = -(10\text{ V/m}^2)w\ell = -(d/dt)\int \mathbf{B}\cdot d\mathbf{A} = -(dB/dt)w\ell$, or $dB/dt = 10$ T/s.

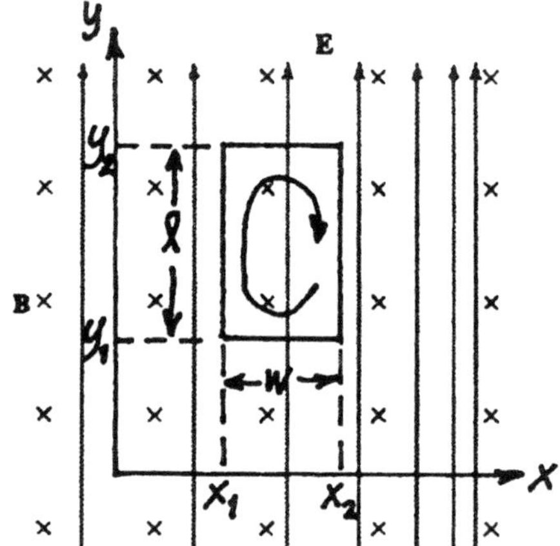

FIGURE 31-54 Problem 39 Solution.

Problem

41. Figure 31-56 shows a magnetic field pointing into the page; the field is confined to a layer of thickness h in the vertical direction but extends infinitely to the left and right. The field strength is increasing with time: $B = bt$, where b is a constant. Find an expression for the electric field at all points outside the field region. *Hint:* Consult Example 30-5.

FIGURE 31-56 Problem 41.

Solution

A changing magnetic field acts as a source for an induced electric field, just like a current density is a source for a magnetic field. In fact, Faraday's law (for a loop fixed in space), $\oint_{\text{loop}} E \cdot d\ell = -d\phi_B/dt = \int_{\text{surface}} (\partial \mathbf{B}/\partial t) \cdot d\mathbf{A}$, is analogous to Ampère's law, $\oint_{\text{loop}} \mathbf{B} \cdot d\ell = \mu_0 I_{\text{encircled}} = \int_{\text{surface}} \mu_0 \mathbf{J} \cdot d\mathbf{A}$ (compare $\mathbf{E}$ and $\partial\mathbf{B}/\partial t$ with $\mathbf{B}$ and $\mu_0\mathbf{J}$). The geometry of the source and symmetry of the field in Fig. 31-56 is similar to that in Fig. 30-25 for an infinite current sheet. The induced electric field, $\mathbf{E}$, should have the same magnitude above and below the source region for $\partial\mathbf{B}/\partial t$, and should circulate in a CCW sense so as to oppose the increase of flux into the page (i.e., CCW circulation is out of the page, opposite to the normal to an area into the page). For the rectangular loop shown added to Fig. 31-56, $\oint_{\text{loop}} \mathbf{E} \cdot d\ell = 2E\ell = \int_{\text{area}} (\partial B/\partial t)\, dA = (\partial B/\partial t)\ell h$, or $E = \frac{1}{2}|\partial B/\partial t|\, h = \frac{1}{2}bh$.

Paired Problems

Problem

43. A magnetic field is given by $\mathbf{B} = B_0(x/x_0)^2\hat{\mathbf{k}}$, where B_0 and x_0 are constants. Find an expression for the magnetic flux through a square of side $2x_0$ that lies in the x-y plane with one corner at the origin and two sides coinciding with the positive x and y axes.

Solution

Take elements of area, $d\mathbf{A} = 2x_0\, dx\hat{\mathbf{k}}$, which are rectangular strips parallel to the y-axis. Then

$$\phi_B = \int_{\text{square}} \mathbf{B} \cdot d\mathbf{A} = \left(\frac{B_0}{x_0^2}\right) 2x_0 \int_0^{2x_0} x^2\, dx$$
$$= (2B_0/x_0)(2x_0)^3/3 = 16B_0x_0^2/3.$$

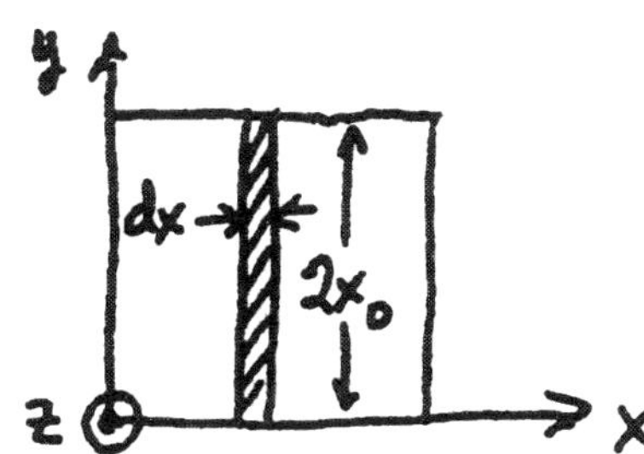

Problem 43 Solution.

Problem

45. A uniform magnetic field is given by $\mathbf{B} = bt\hat{\mathbf{k}}$, where $b = 0.35$ T/s. Find the current in a conducting loop with area 240 cm^2 and resistance 0.20 Ω that lies in the x-y plane. In what direction is the current, as viewed from the positive z-axis?

Solution

A normal to the loop parallel to the z-axis corresponds to CCW positive circulation (via the right-hand rule), when viewed *from* the positive z-axis. Faraday's and Ohm's law give the current in the loop: $I = \mathcal{E}/R = -(d\phi_B/dt)/R = -A(dB/dt)/R = -Ab/R = -(240\text{ cm}^2)(0.35\text{ T/s})/(0.2\ \Omega) = -42.0$ mA, the minus sign indicating a CW circulation viewed from the positive z-axis.

Problem

47. A pair of vertical conducting rods are a distance ℓ apart and are connected at the bottom by a resistance R. A conducting bar of mass m runs horizontally between the rods and can slide freely down them while maintaining electrical contact. The whole apparatus is in a uniform magnetic field $\mathbf{B}$ pointing horizontally and perpendicular to the bar. When the bar is released from rest it soon reaches a constant speed. Find this speed.

Solution

When the bar is falling, a motional emf causes an induced current to flow in the bar (in the direction of $\mathbf{v} \times \mathbf{B}$) as shown. When we are looking horizontally in the direction of $\mathbf{B}$ (into the page), the forces on the bar are gravity, mg downward, and the magnetic force, $I\ell B$ upward, where the induced current opposes the decrease in flux. The velocity is constant when $I\ell B = mg$. Now, $\phi_B = B\ell y$ and $I = \mathcal{E}/R = -(d\phi_B/dt)/R = -(B\ell/R)(dy/dt) = B\ell v/R$ (where $dy/dt = -v$ is the speed downward). Therefore, $v = IR/B\ell = mgR/B^2\ell^2$.

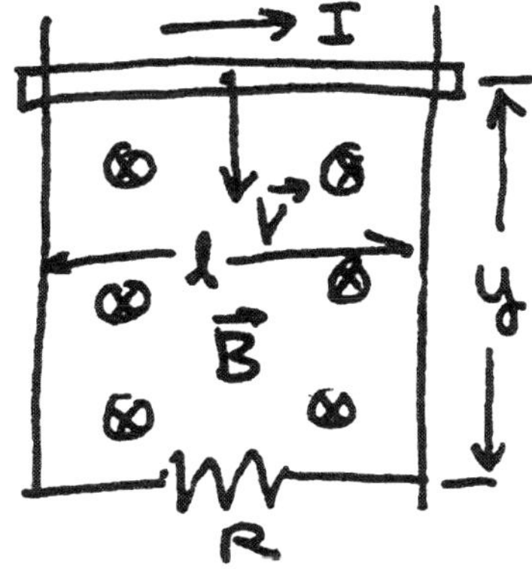

Problem 47 Solution.

Problem

49. Figure 31-58 shows an unusual design for a generator, consisting of a conducting bar that

rotates about a central axis while making contact with a conducting ring of radius R. A uniform magnetic field is perpendicular to the ring. Wires from the axis and ring carry power to a load. Find an expression for the emf induced in this generator when the bar rotates with angular speed ω.

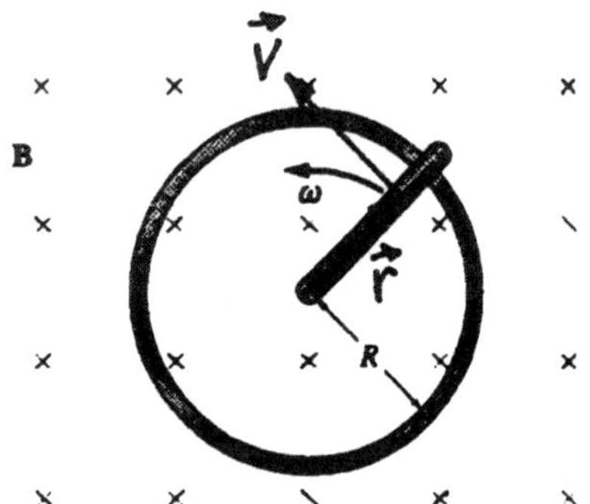

FIGURE 31-58 Problem 49.

Solution

Each rotation, the bar sweeps through an area of πR^2 perpendicular to the magnetic field B, so the flux changes by $\Delta\phi_B = \pi R^2 B$ in one period of rotation $\Delta t = 2\pi/\omega$. Then the magnitude of the induced emf is $\mathcal{E} = |-\Delta\phi_B/\Delta t| = \pi R^2 B/(2\pi/\omega) = \frac{1}{2}\omega R^2 B$. Alternatively, at a point r on the bar, there is a motional emf resulting from an equivalent electric field of $\mathbf{v}\times\mathbf{B} = -vB\hat{\mathbf{r}} = -\omega rB\hat{\mathbf{r}}$ (minus $\hat{\mathbf{r}}$ is toward the center). The emf developed across the length of the bar is $\mathcal{E} = \int_0^R \mathbf{E}\cdot d\mathbf{r} = -\omega B\int_0^R r\,dr = -\frac{1}{2}\omega BR^2$ (the axis is positive relative to the rim).

Problem

51. An electron is inside a solenoid, 28 cm from the solenoid axis. It experiences an electric force of magnitude 1.3 fN. At what rate is the solenoid's magnetic field changing?

Solution

The electric field has magnitude $E = F/e$. If we suppose that this is the electric field induced by the changing magnetic field in the solenoid, and use the axial symmetric approximation in Example 31-9 (modified as in the accompanying exercise, $\phi_B = \pi r^2 B$ inside the solenoid, and $2\pi rE = -\pi r^2(dB/dt)$), then $E = \frac{1}{2}r\,|dB/dt| = F/e$, or $|dB/dt| = 2F/re =$ $2(1.3\text{ fN})/(28\text{ cm}\times 1.6\times 10^{-19}\text{ C}) = 58.0\text{ T/ms}$.

Supplementary Problems

Problem

53. At time prior to $t = 0$, there is no current in either the solenoid or the small coil of Problem 19. Subsequently, the current in the small coil is observed to increase at 10 μA/s. What is the solenoid current as a function of time?

Solution

If we assume that the current in the small coil is only the current induced by the solenoid, then differentiation of the result of Problem 19 and substitution of numerical data gives

$$10\frac{\mu\text{A}}{\text{s}} = \frac{dI_C}{dt} = \mu_0\left(\frac{N_S N_C}{\ell}\right)\frac{1}{4}\pi D_S^2\frac{1}{R}\frac{d^2 I_S}{dt^2},$$

or

$$\frac{d^2 I_S}{dt^2} = 1.62\frac{\text{A}}{\text{s}^2}.$$

(We only need to find the magnitude of I_S, since its direction was discussed in Problem 19.) The solution of this differential equation, which satisfies the initial conditions ($I_S = 0$ and $dI_S/dt \sim I_C = 0$ at $t = 0$), is simply $I_S = \frac{1}{2}(1.62\text{ A/s}^2)t^2$.

Problem

55. So-called magnetohydrodynamic generators have been proposed as a means of extracting electrical energy from charged particles released in fusion reactions; they've also been suggested as a way to generate electricity from flowing water. An MHD generator consists of two metal plates on either side of a channel carrying conducting fluid in a magnetic field, as shown in Fig. 31-60. The magnetic force on free charges in the fluid drives positive charge to one plate, negative to the other. If there's no electrical load connected across the plates, the electric field that develops eventually halts any further charge motion. (a) Show in this

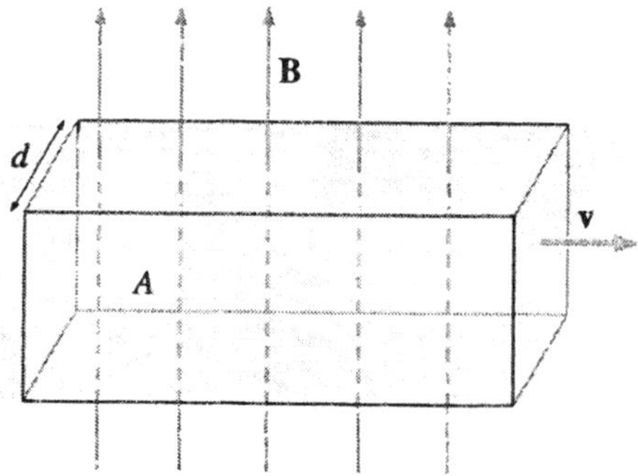

FIGURE 31-60 Problem 55.

case that the voltage between the plates is $V = vBd$, where v is the fluid velocity and d the plate spacing. (b) Now suppose a resistance R is connected between the plates. Show that the current through R is $I = \frac{vABd}{\rho d+AR}$, where A is the plate area and ρ is the resistivity of the fluid.

Solution

(a) The equivalent motional electric field on charge carriers in the fluid channel, $\mathbf{E}_i = \mathbf{F}_{\text{mag}}/q = \mathbf{v} \times \mathbf{B}$, with magnitude $E_i = vB$, results in an emf of $\mathcal{E}_i = \int \mathbf{E}_i \cdot d\boldsymbol{\ell} = vBd$ between the plates. When no circuit is present, the static field of accumulated charge cancels the motional emf, and the potential difference between the plates has the same magnitude. (b) When current is free to flow in a circuit containing the plates, its resistance outside the fluid channel is R, and for the volume of conducting fluid inside the channel, is $\rho d/A$ (assuming equipotential surfaces parallel to the plates, no fringing fields, etc.). Then the magnitude of the current generated by the motional emf in part (a) is $I = \mathcal{E}_i/R_{\text{tot}} = vBd/(R + \rho d/A)$, similar to a battery of internal resistance $\rho d/A$, and emf vBd, connected to a resistance R.

Problem

57. Clever farmers whose lands are crossed by large power lines have been known to steal power by stringing wire near the power line and making use of the induced current—a practice that has been ruled legally to be theft. The scene of a particular crime is shown in Fig. 31-61. The power line carries 60-Hz alternating current with a peak current of 10 kA (that is, the current is given by $I = I_0 \sin \omega t$, where $I_0 = 10$ kA and $\omega = 2\pi f$, with $f = 60$ Hz). (a) If the farmer wants a peak voltage of 170 V, what should be the length ℓ of the loop shown in Fig. 31-61? (170 V is the peak of standard 120-V AC power.) (b) If all the equipment the farmer connects to the loop has an equivalent resistance of 5.0 Ω, what is the farmer's average power consumption? *Note:* The *average* power consumption is half the product of the peak voltage and peak current. (c) If the power company charges 10¢ per kWh, what is the monetary value of the energy stolen each day? (d) Without examining the farmer's lands, how, in principle, could the power company know that a crime is being committed?

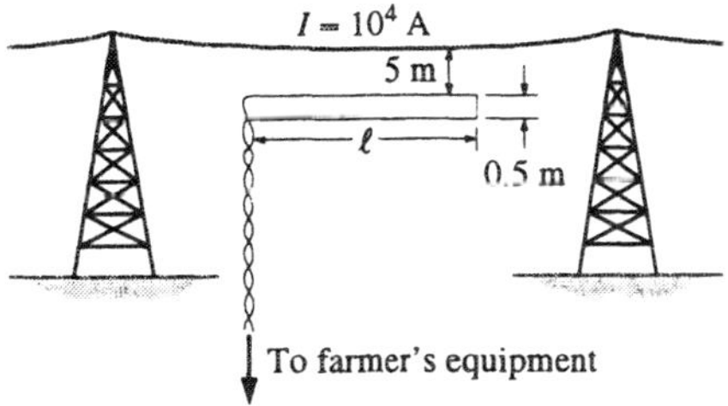

FIGURE 31-61 Problem 57.

Solution

(a) The induced emf $\mathcal{E}$ in the farmer's loop (which depends on the dimensions of the loop) can be calculated from the flux found in Example 31-2 for a similar configuration of wire and loop, $\phi_B = (\mu_0/2\pi) \times I\ell \ln(1 + w/a)$. (The dimensions of the loop and its distance from the power line are shown in Fig. 31-61.) For $I = I_0 \sin \omega t$ (and $dI/dt = \omega I_0 \cos \omega t$), Faraday's law gives $\mathcal{E} = -d\phi_B/dt = -(\mu_0/2\pi)\omega I_0 \ell \cos \omega t \ln \times (1 + w/a) \equiv -\mathcal{E}_P \cos \omega t$, where $\mathcal{E}_P$ is the peak voltage in the farmer's loop. Using the numerical data given, we find:

$$\ell = \frac{170 \text{ V}}{(2\times10^{-7} \text{ N/A}^2)(10^4 \text{ A})(2\pi\times60 \text{ Hz}) \ln(5.5/5)} = 2.37 \text{ km}.$$

(b) For sinusoidal current, the average power is one half the peak power (see Chapter 33), so the average power stolen is $\mathcal{P}_{\text{av}} = \frac{1}{2}\mathcal{E}_P^2/R = \frac{1}{2}(170 \text{ V})^2/5 \text{ Ω} = 2.89$ kW. (c) For continuous consumption, the farmer's dishonesty costs (2.89 kW)(24 h/d) (\$0.10/kW·h) = \$6.94/d. (d) The company could measure the power on either side of the farmer's property to detect the difference, but, the stolen power is such a small fraction of the total power output, very accurate measurements would be necessary. Certainly, the company is producing more power than it's being paid for, and this is, in principle, detectable.

Problem

59. A conducting disk with radius a, thickness h, and resistivity ρ is inside a solenoid of circular cross section. The disk axis coincides with the solenoid axis. The magnetic field in the solenoid is given by $B = bt$, with b a constant. Find expressions for (a) the current density in the disk as a function of the distance r from the disk center and (b) the rate of power dissipation in the entire disk. *Hint:* Consider the disk to be made up of infinitesimal conducting loops.

Solution

The changing magnetic field in the solenoid will induce a circular current density J in the disk. (a) From the exercise accompanying Example 31-9 (as mentioned in the solutions to Problems 37, 42, and 51) the induced electric field in the disk has magnitude $E = \frac{1}{2}rb$ (when $dB/dt = b$), so Ohm's law (in point form) gives $J = E/\rho = rb/2\rho$. (b) The result of Problem 27-73 gives the power dissipated per unit volume of disk (the joule heat). For volume elements, take rings of radius r, thickness dr, and width h, so

$$\mathcal{P}_J = \int J^2 \rho dV = \frac{b^2}{4\rho}\int_0^a r^2 2\pi rh\, dr = \frac{b^2}{4\rho}\cdot 2\pi h\cdot\frac{a^4}{4} = \frac{\pi b^2 a^4 h}{8\rho}.$$

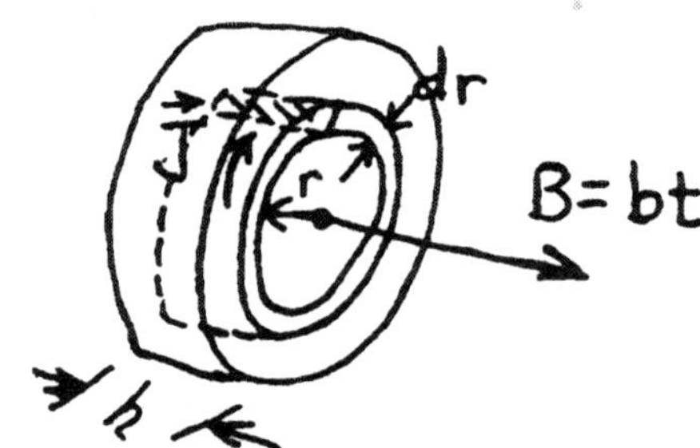

Problem 59 Solution.

Problem

61. Find an expression for the speed of the left-hand resistor in Problem 32 as a function of time, in terms of its mass m, the field strength B, the speed v of the right-hand bar, the time t, and the resistance R_{left} and R_{right}.

Solution

When the right-hand resistor has constant speed v, the left-hand resistor (initially at rest at $t = 0$) will begin moving to the right, as explained in the solution to Problem 32. We assume that the only horizontal force acting on the resistor is the magnetic force, $F_B = I\ell B$ (to the right), so its horizontal equation of motion is $m\, dv_{\text{left}}/dt = I\ell B$. The current has magnitude $I = (1/R_{\text{tot}})(d\phi_B/dt) = (\ell B/R_{\text{tot}})\, d(x_{\text{right}} - x_{\text{left}})/dt = (\ell B/R_{\text{tot}})(v - v_{\text{left}})$, where $R_{\text{tot}} = R_{\text{left}} + R_{\text{right}}$, and x and $v = dx/dt$ are the position and speed of a resistor. Newton's second law becomes $m\, dv_{\text{left}}/dt = (B^2\ell^2/R_{\text{tot}})(v - v_{\text{left}})$. This equation is easily solved by separation of variables and integration:

$$\int \frac{dv_{\text{left}}}{v - v_{\text{left}}} = \int \left(\frac{B^2\ell^2}{mR_{\text{tot}}}\right) dt = -\ln\left(1 - \frac{v_{\text{left}}}{v}\right) = \left(\frac{B^2\ell^2}{mR_{\text{tot}}}\right) t, \text{ or } v_{\text{left}} = v(1 - e^{B^2\ell^2 t/mR_{\text{tot}}}).$$

Problem

63. A *flip coil* consists of a small coil used to measure magnetic fields. The flip coil is placed in a magnetic field with its plane perpendicular to the field, and then rotated abruptly through 180° about an axis in the plane of the coil. The coil is connected to instrumentation to measure the total charge Q that flows during this process. If the coil has N turns of area A and if its rotation axis is perpendicular to the magnetic field, show that the field strength is given by $B = QR/2NA$, where R is the coil resistance.

Solution

Initially, the flux through the flip coil is $\phi_B = NBA$, but is reversed to $-NBA$ when the coil is rotated 180°, so $\Delta\phi_B = -2NBA$. The total charge which flows is $\Delta Q = I_{\text{av}}\Delta t$, where I_{av} is the average induced current and Δt the time for the rotation. From Faraday's and Ohm's laws, $I_{\text{av}} = (-\Delta\phi_B/\Delta t)/R = 2NBA/R\,\Delta t$, hence $\Delta Q = 2NBA/R$. If the properties of the coil are known and the total charge is measured, one can find the magnetic field strength, $B = R\,\Delta Q/2NA$.

CHAPTER 32 INDUCTANCE AND MAGNETIC ENERGY

ActivPhysics can help with these problems: Activity 14.1

Section 32-1: Mutual Inductance

Problem

1. Two coils have a mutual inductance of 2.0 H. If current in the first coil is changing at the rate of 60 A/s, what is the emf in the second coil?

Solution

From Equation 32-2, $\mathcal{E}_2 = -M(dI_1/dt) = -(2\text{ H})\times(60\text{ A/s}) = -120\text{ V}$. (The minus sign, Lenz's law, signifies that an induced emf opposes the process which creates it.)

Problem

3. The current in one coil is given by $I = I_p \sin 2\pi ft$, where $I_p = 75$ mA, $f = 60$ Hz, and $t =$ time. Find the peak emf in a second coil if the mutual inductance between the coils is 440 mH.

Solution

Suppose $I_1 = I_p \sin 2\pi ft$ in Equation 32-2. Then $\mathcal{E}_2 = -M\ dI_1/dt = -2\pi fMI_p \cos 2\pi ft$, and the peak emf (when $\cos 2\pi ft = \pm 1$) is $2\pi fMI_p = (2\pi\times 60\text{ Hz})(440\text{ mH})(75\text{ mA}) = 12.4\text{ V}$.

Problem

5. An alternating current given by $I_p \sin 2\pi ft$ is supplied to one of two coils whose mutual inductance is M. (a) Find an expression for the emf in the second coil. (b) When $I_p = 1.0$ A and $f = 60$ Hz, the peak emf in the second coil is measured at 50 V. What is the mutual inductance?

Solution

(a) From Equation 32-2, $\mathcal{E}_2 = -M\ dI_1/dt = -M2\pi fI_p \cos(2\pi ft)$. (b) The peak value of the cosine is 1, so $|M| = \mathcal{E}_{2p}/2\pi fI_p = 50\text{ V}/(2\pi\times 60\text{ Hz})(1\text{ A}) = 133\text{ mH}$. (From the information given, only the magnitude of M can be determined; its sign depends on how the coils are coupled.)

Problem

7. Two long solenoids of length ℓ both have n turns per unit length. They have circular cross sections with radii R and $2R$, respectively. The smaller solenoid is mounted inside the larger one, with their axes coinciding. Find the mutual inductance of this arrangement, neglecting any nonuniformity in the magnetic field near the ends.

Solution

All of the flux from the smaller solenoid (number one) links the larger solenoid (number two), so $\phi_{B,2} = N_2\mathbf{B}_1\cdot\mathbf{A}_1 = N_2(\mu_0 n_1 I_1)(\pm\pi R^2) = \pm\mu_0 n^2\pi R^2\ell I_1$, since both solenoids have the same number of turns and length. Dividing by I_1 gives M (see Equation 32-1). Note that the sign of M depends on the relative direction of the windings in the two solenoids. (Alternatively, only a fraction A_1/A_2 of the flux from the larger solenoid links the smaller solenoid, so $\phi_{B,1} = \pm N_1B_2A_1$ and $M = \phi_{B,1}/I_2$ is the same.)

Problem

9. A rectangular loop of length ℓ and width w is located a distance a from a long, straight, wire, as shown in Fig. 32-20. What is the mutual inductance of this arrangement?

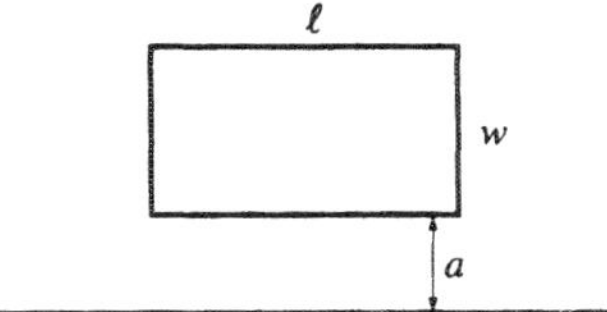

FIGURE 32-20 Problem 9.

Solution

When current I_1 flows to the left in the wire, the flux through the loop is $\phi_{B,2} = (\mu_0 I_1/2\pi)\int_a^{a+w}\ell\ dr/r = (\mu_0 I_1\ell/2\pi)\ln(1+w/a)$ (see Example 31-2). Then Equation 32-1 gives $M = \phi_{B,2}/I_1 = (\mu_0\ell/2\pi)\times\ln(1+w/a)$. (In calculating the flux, the normal to the loop area was taken into the page, so the positive sense of circulation around the loop is CW. This determines the direction of the induced emf $\mathcal{E}_2$ in Equation 32-2.)

Section 32-2: Self-Inductance

Problem

11. What is the self-inductance of a solenoid 50 cm long and 4.0 cm in diameter that contains 1,000 turns of wire?

Solution

Equation 32-4 gives $L = \mu_0 N^2 A/\ell = (4\pi\times 10^{-7}\,\text{H/m})\times (10^3)^2\pi(2\text{ cm})^2/(50\text{ cm}) = 3.16$ mH. (The long thin solenoid approximation is valid here.)

Problem

13. A 2.0-A current is flowing in a 20-H inductor. A switch is opened, interrupting the current in 1.0 ms. What emf is induced in the inductor?

Solution

Assume that the current changes uniformly from 2 A to zero in 1 ms (or consider average values). Then $dI/dt = -2$ A/ms, and Equation 32-5 gives $\mathcal{E} = -(20\text{ H})(-20\text{ A/ms}) = 40$ kV. (The emf opposes the decreasing current.)

Problem

15. A cardboard tube measures 15 cm long by 2.2 cm in diameter. How many turns of wire must be wound on the full length of the tube to make a 5.8-mH inductor?

Solution

From Equation 32-4, $N = \sqrt{L\ell/\mu_0 A} = [(5.8\text{ mH})\times (15\text{ cm})/(4\pi\times 10^{-7}\text{ H/m})\pi(1.1\text{ cm})^2]^{1/2} = 1.35\times 10^3$ turns.

Problem

17. The emf in a 50-mH inductor has magnitude $|\mathcal{E}| = 0.020t$, with t in seconds and $\mathcal{E}$ in volts. At $t = 0$ the inductor current is 300 mA. (a) If the current is increasing, what will be its value at $t = 3.0$ s? (b) Repeat for the case when the current is decreasing.

Solution

(a) $\mathcal{E}$ has the opposite sign to dI/dt in Equation 32-5. When I is increasing, $dI/dt > 0$, $\mathcal{E}$ is negative, $\mathcal{E} = -|\mathcal{E}|$. Thus,

$$\frac{dI}{dt} = -\frac{\mathcal{E}}{L} = \frac{(0.02\text{ V/s})t}{(0.05\text{ H})} = \left(0.4\frac{\text{A}}{\text{s}^2}\right)t,$$

and

$$I = \frac{1}{2}\left(0.4\frac{\text{A}}{\text{s}^2}\right)t^2 + I_0.$$

At $t = 3$ s, $I = (0.2\text{ A/s}^2)(3\text{ s})^2 + 0.3\text{ A} = 2.1$ A. (b) For $dI/dt < 0$, $\mathcal{E} = |\mathcal{E}| > 0$, so $dI/dt = -(0.4\text{ A/s}^2)t$, and $I = -(0.2\text{ A/s}^2)t^2 + I_0$. At $t = 3$ s in this case, $I = -1.5$ A. (After $t = \sqrt{3/2}$ s, the current reverses direction and begins increasing in absolute value.)

Problem

19. A 2,000-turn solenoid is 65 cm long and has cross-sectional area 30 cm^2. What rate of change of current will produce a 600-V emf in this solenoid?

Solution

The self-inductance (of this long, thin solenoid) is $L = \mu_0 N^2 A/\ell = (4\pi\times 10^{-7}\text{ H/m})(2000)^2(30\text{ cm}^2)\div(65\text{ cm}) = 23.2$ mH (see Equation 32-4), so Equation 32-5 gives $|dI/dt| = |\mathcal{E}|/L = (600\text{ V})\div(23.2\text{ mH}) = 25.9$ A/ms.

Problem

21. The emf in a 50-mH inductor is given by $\mathcal{E} = \mathcal{E}_p \sin\omega t$, where $\mathcal{E}_p = 75$ V and $\omega = 140\text{ s}^{-1}$. What is the peak current in the inductor? (Assume the current swings symmetrically about zero.)

Solution

From Equation 32-5, $dI/dt = -(\mathcal{E}_p/L)\sin\omega t$, so integration yields $I(t) = (\mathcal{E}_p/\omega L)\cos\omega t$. (Since $I(t)$ is symmetric about $I = 0$, the constant of integration is zero.) The peak current is $I_p = \mathcal{E}_p/\omega L = 75\text{ V}/(140\text{ s}^{-1}\times 50\text{ mH}) = 10.7$ A.

Section 32-3: Inductors in Circuits

Problem

23. Show that the inductive time constant has the units of seconds.

Solution

A henry is a volt· second/ampere (see Equation 32-5), so the units of $\tau_L = L/R$ are H/Ω = V·s/Ω·A = s.

Problem

25. The current in a series RL circuit rises to 20% of its final value in 3.1 μs. If $L = 1.8$ mH, what is the resistance R?

Solution

The buildup of current in an RL circuit with a battery is given by Equation 32-8, $I(t) = I_\infty(1 - e^{-Rt/L})$, where $I_\infty = \mathcal{E}_0/R$ is the final current. Solving for R, one finds $R = -(L/t)\ln(1 - I/I_\infty) = -(1.8\text{ mH}\div 3.1\ \mu\text{s})\ln(1 - 20\%) = 130\ \Omega$.

Problem

27. A 10-H inductor is wound of wire with resistance 2.0 Ω. If the inductor is connected across an ideal 12-V battery, how long will it take the current to reach 95% of its final value?

Solution

Reference to the solution to Problem 25 shows that $t = -(L/R)\ln(1 - I/I_\infty) = -(10\text{ H}/2\ \Omega)\ln(1-0.95) =$ 15.0 s. (The percentage I/I_∞ is independent of $\mathcal{E}_0$.)

Problem

29. In Fig. 32-8*a*, take $R = 2.5$ kΩ and $\mathcal{E}_0 = 50$ V. When the switch is closed, the current through the inductor rises to 10 mA in 30 μs. (a) What is the inductance? (b) What will be the current in the circuit after many time constants?

Solution

(b) After a long time ($t \to \infty$), the exponential term in Equation 32-8 is negligible. Thus, $I_\infty = \mathcal{E}_0/R =$ 50 V/2.5 kΩ = 20 mA. (a) The current has risen to half its final value in 30 μs. Thus (Equation 32-8 again), $\frac{1}{2} = 1 - e^{-Rt/L}$, or $L = Rt/\ln 2 = (2.5\text{ k}\Omega)\times(30\ \mu\text{s})/\ln 2 = 108$ mH.

Problem

31. In Fig. 32-8a, take $R = 100\ \Omega$, $L = 2.0$ H, and $\mathcal{E}_0 = 12$ V. At 20 ms after the switch is closed, what are (a) the circuit current, (b) the inductor emf, (c) the resistor voltage, (d) the rate of change of the circuit current, and (e) the power dissipation in the resistor?

Solution

(a) The time constant is $L/R = 2\text{ H}/100\ \Omega = 20$ ms, and the final current is $\mathcal{E}_0/R = 12\text{ V}/100\ \Omega = 120$ mA. After one time constant, Equation 32-8 gives $I =$ 120 mA$(1 - e^{-1}) = 75.9$ mA. (b) The voltage drop across the inductor is (from Equation 32-7) $V_L = \mathcal{E}_0 - IR = -\mathcal{E}_L = \mathcal{E}_0 e^{-Rt/L}$. After one time constant, $V_L = (12\text{ V})e^{-1} = 4.41$ V. (Note that $V_L + V_R = \mathcal{E}_0$ is Kirchhoff's loop law.) (c) $V_R = IR = (75.9\text{ mA})\times(100\ \Omega) = 7.59$ V. (Alternatively, $V_R = \mathcal{E}_0 - V_L =$ 12 V − 4.41 V.) (d) From Equation 32-5 and the loop law, $V_L = -\mathcal{E}_L = L\,dI/dt$. After one time constant, $dI/dt = V_L/L = 4.41\text{ V}/2\text{ H} = 2.21$ A/s. (e) $P_R = I^2R = (75.9\text{ mA})^2(100\ \Omega) = 575$ mW.

Problem

33. Resistor R_2 in Fig. 32-22 is to limit the emf that develops when the switch is opened. What should be its value in order that the inductor emf not exceed 100 V?

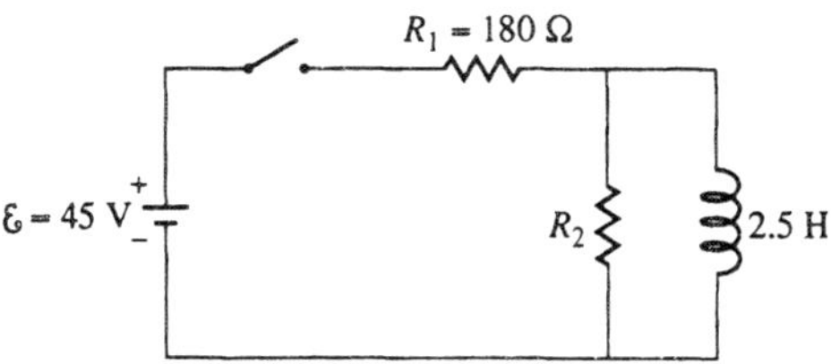

FIGURE 32-22 Problem 33.

Solution

As explained in Example 32-6, when the switch is opened (after having been closed a long time), the voltage across R_2 (which equals the inductor emf) is $V_2 = I_2R_2 = \mathcal{E}_0R_2/R_1$. If we choose to limit this to no more than 100 V, then $R_2 \le (100\text{ V})(180\ \Omega)/45\text{ V} =$ 400 Ω.

Problem

35. A 5.0-A current is flowing through a nonideal inductor with $L = 500$ mH. If the inductor is suddenly short-circuited, the inductor current drops to 2.5 A in 6.9 ms. What is the resistance of the inductor?

Solution

A real inductor can be represented by a resistance, R, in series with an inductance, L. When short-circuited, the inductor constitutes an LR circuit without a battery, and the current decays exponentially with time constant L/R, $I = I_0e^{-Rt/L}$ (see Equation 32-9). When the given data is substituted, we can solve for the resistance,

$$R = \left(\frac{L}{t}\right)\ln\left(\frac{I_0}{I}\right) = \left(\frac{500\text{ mH}}{6.9\text{ ms}}\right)\ln 2 = 50.2\ \Omega.$$

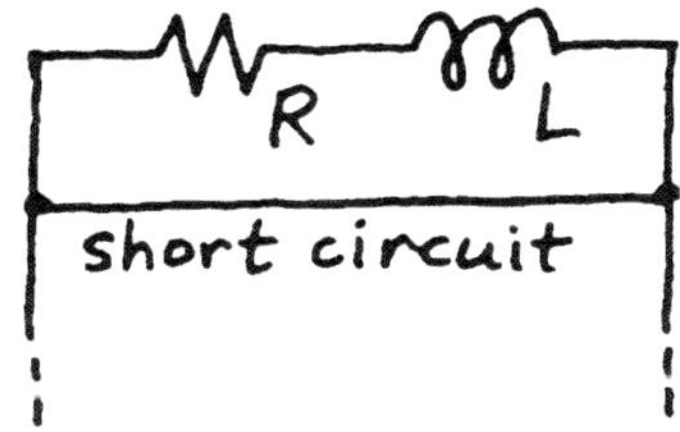

Problem 35 Solution.

Problem

37. In Fig. 32-24, take $\mathcal{E}_0 = 20$ V, $R_1 = 10\ \Omega$, $R_2 = 5.0\ \Omega$, and assume the switch has been open

for a long time. (a) What is the inductor current immediately after the switch is closed? (b) What is the inductor current a long time after the switch is closed? (c) If after a long time the switch is again opened, what will be the voltage across R_1 immediately afterward?

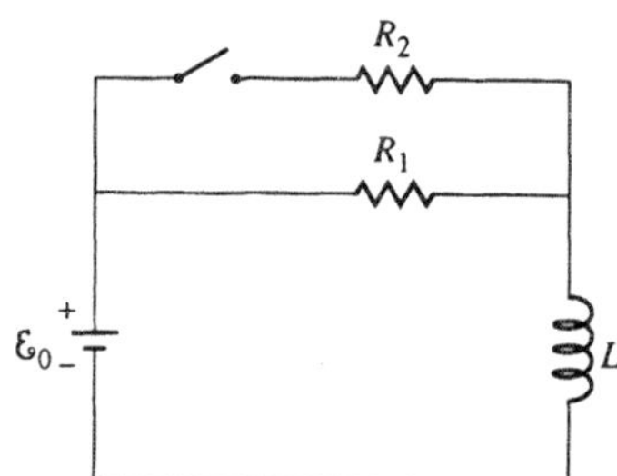

FIGURE 32-24 Problem 37.

Solution

(a) If the switch has been open a long time, a steady current flows through the inductance ($dI_L/dt = 0$). When the switch is closed (at $t = 0$), I_L cannot change instantaneously, so $I_L(0) = \mathcal{E}/R_1 = 20\ \text{V}/10\ \Omega = 2$ A. (Of course, $I_1(0) = I_L(0)$, and $I_2(0) = 0$.) (b) After another long time ($t \to \infty$), the currents are steady again and $\mathcal{E}_L = 0$ (the inductance behaves like a short circuit). The resistors are in parallel; therefore $I_L(\infty) = \mathcal{E}(1/R_1 + 1/R_2) = 20\ \text{V}(\frac{1}{5} + \frac{1}{10})\ \Omega^{-1} = 6$ A. (c) When the switch is again opened, the current through R_2 is zero, but I_L cannot change instantly, so $I_L = I_1 = I_L(\infty) = 6$ A. Thus, the voltage across R_1 is $V_1 = I_1R_1 = (6\ \text{A})(10\ \Omega) = 60$ V.

Section 32-4: Magnetic Energy

Problem

39. What is the current in a 10-mH inductor when the stored energy is 50 μJ?

Solution

From Equation 32-10, $I = \sqrt{2U/L} = \sqrt{2(50\ \mu\text{J})/10\ \text{mH}} = 0.1$ A.

Problem

41. A 12-V battery, 5.0-Ω resistor, and 18-H inductor are connected in series and allowed to reach a steady state. (a) What is the energy stored in the inductor? (b) Once in the steady state, over what time interval is the energy dissipated in the resistor equal to that stored in the inductor?

Solution

(a) The steady state (i.e., final) current in an RL circuit with emf $\mathcal{E}_0$ is $I = \mathcal{E}_0/R$, so the energy stored in the inductor (Equation 32-10) is $U_L = \frac{1}{2}LI^2 = \frac{1}{2}(18\ \text{H})(12\ \text{V}/5\ \Omega)^2 = 51.8$ J. (b) Energy is dissipated in the resistor at the rate $\mathcal{P}_R = I^2R$ (the joule heat), so the time interval queried is $\Delta t = U_L/\mathcal{P}_R = \frac{1}{2}LI^2/I^2R = L/2R = 18\ \text{H}/(2{\times}5\ \Omega) = 1.8$ s.

Problem

43. The current in a 2.0-H inductor is decreased linearly from 5.0 A to zero over 10 ms. (a) What is the average rate at which energy is being extracted from the inductor during this time? (b) Is the instantaneous rate constant?

Solution

(a) The energy falls from $U_i = \frac{1}{2}LI^2 = \frac{1}{2}(2\ \text{H})\times(5\ \text{A})^2 = 25$ J to $U_f = 0$ in $\Delta t = 10$ ms, so the rate of decrease is $\Delta U/\Delta t = -25\ \text{J}/10\ \text{ms} = -2.5$ kW.
(b) The discussion in the text leading to Equation 32-10 shows that the instantaneous power is $\mathcal{P}_L = LI(dI/dt)$, so even if dI/dt is constant, I and $\mathcal{P}_L$ are not.

Problem

45. The current in a 2.0-H inductor is increasing. At some instant, the current is 3.0 A and the inductor emf is 5.0 V. At what rate is the inductor's magnetic energy increasing at this instant?

Solution

The rate at which energy is stored in an inductor is $\mathcal{P}_L = LI(dI/dt)$ (see the discussion of "Magnetic Energy in an Inductor" leading to Equation 32-10). When the current is increasing, as for this inductor, $L(dI/dt) = |\mathcal{E}_L|$, and $\mathcal{P}_L = I\,|\mathcal{E}_L| = (3\ \text{A})(5\ \text{V}) = 15$ W.

Problem

47. A superconducting solenoid with inductance $L = 3.5$ H carries 1.8 kA. Copper is embedded in the coils to carry the current in the event of a quench (see Example 32-7). (a) What is the magnetic energy in the solenoid? (b) What is the maximum resistance of the copper that will limit the power dissipation to 100 kW immediately after a loss of superconductivity? (c) With this resistance, how long will it take the power to drop to 50 kW?

Solution

(a) In its superconducting state, the solenoid's stored energy (Equation 32-10) is $U = \frac{1}{2}LI^2 = \frac{1}{2}(3.5 \text{ H})\times(1.8 \text{ kA})^2 = 5.67$ MJ. (b) The current in an inductor cannot change instantaneously, so if the power dissipated in the copper (I^2R) just after a sudden loss of superconductivity must be less than 100 kW, then $R \le 100 \text{ kW}/(1.8 \text{ kA})^2 = 30.9 \text{ m}\Omega$. (c) The power drops to one half its maximum original value, when the current drops to $1/\sqrt{2}$ times its initial value. From Equation 32-9, the decay time is $t = (L/R)\ln(I_0/I) = (3.5 \text{ H}/30.9 \text{ m}\Omega)\ln\sqrt{2} = 39.3$ s.

Problem

49. The Alcator fusion experiment at MIT has a 50-T magnetic field. What is the magnetic energy density in Alcator?

Solution

From Equation 32-11, $u_B = (50 \text{ T})^2/(8\pi\times 10^{-7} \text{ N/A}^2) = 995 \text{ MJ/m}^3$. (This is about 2.8% of the energy density content of gasoline; see Appendix C.)

Problem

51. The magnetic field of a neutron star is about 10^8 T. How does the energy density in this field compare with the energy density stored in (a) gasoline and (b) pure uranium-235 (mass density $19\times 10^3 \text{ kg/m}^3$)? Consult Appendix C.

Solution

The energy density in a field of this strength is $u_B = B^2/2\mu_0 = (10^8 \text{ T})^2/(8\pi\times 10^{-7} \text{ H/m}) = 3.98\times 10^{21} \text{ J/m}^3$ (see Equation 32-11). This is about (a) 1.1×10^{11} times the energy density content of gasoline ($44 \text{ MJ/kg} \times 800 \text{ kg/m}^3 = 3.52\times 10^{10} \text{ J/m}^3$), and (b) 2600 times that of pure U^{235} ($8\times 10^{13} \text{ J/kg} \times 19\times 10^3 \text{ kg/m}^3 = 1.52\times 10^{18} \text{ J/m}^3$).

Problem

53. A single-turn loop of radius R carries current I. How does the magnetic energy density at the loop center compare with that of a long solenoid of the same radius, carrying the same current, and consisting of n turns per unit length?

Solution

The energy density at the center of the loop is $u_B^{(\text{loop})} = B^2/2\mu_0 = (\mu_0 I/2R)^2/2\mu_0 = \mu_0 I^2/8R^2$ (see Equations 30-3 and 32-11). In a long thin solenoid of the same radius, $u_B^{(\text{solenoid})} = (\mu_0 nI)^2/2\mu_0 = \mu_0 n^2 I^2/2$, so the ratio of $u_B^{(\text{loop})}$ to $u_B^{(\text{solenoid})}$ is $1/4n^2R^2$.

Problem

55. A toroidal coil has inner radius R and a square cross section of side ℓ (Fig. 32-25). It is wound with N turns of wire, and carries a current I. Show that the magnetic energy in the toroid is given by

$$U = \frac{\mu_0 N^2 I^2 \ell}{4\pi}\ln\left(\frac{R+\ell}{R}\right).$$

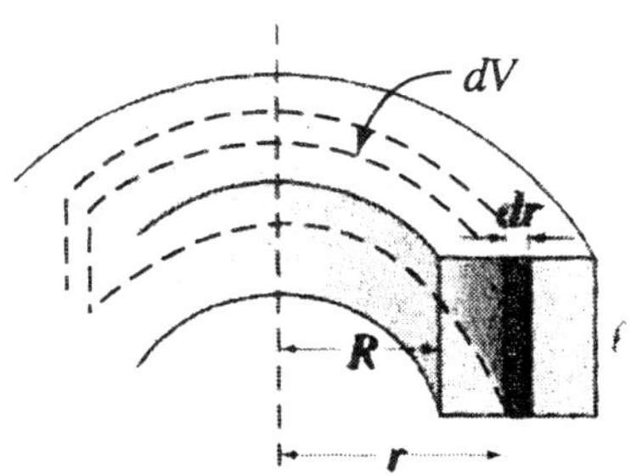

FIGURE 32-25 Problem 55.

Solution

The magnetic field inside a cylindrically symmetric toroid was found by using Ampere's law, $B = \mu_0 NI \div 2\pi r$, where r is the distance from the central axis (see Equation 30-12). The energy in this magnetic field can be found by integrating the energy density, $u_B = B^2/2\mu_0$, over the volume of the toroid, with volume elements consisting of cylindrical shells of radius r, appropriate height, and thickness dr. For a toroid of square cross section of side ℓ, and inner radius R, $dV = 2\pi r\ell\, dr$ (see figure), and

$$U = \int_{\text{toroid}} u_B\, dV = \int_R^{R+\ell} \frac{1}{2\mu_0}\left(\frac{\mu_0 NI}{2\pi r}\right)^2 2\pi r\ell\, dr$$
$$= \frac{\mu_0 N^2 \ell I^2}{4\pi}\int_R^{R+\ell}\frac{dr}{r} = \frac{\mu_0 N^2 \ell I^2}{4\pi}\ln\left(1+\frac{\ell}{R}\right).$$

An alternative approach is to use the flux linkage of the toroid from the solution to Problem 31-30, $N\phi_B = \phi_{\text{toroid}} = LI$ (recall that $a = R$, $b = R+\ell$, and ϕ_B was the flux through one turn), and $U = \frac{1}{2}I\phi_{\text{toroid}}$.

Paired Problems

Problem

57. Two coils have mutual inductance M. The current supplied to coil A is given by $I = bt^2$. Find an expression for the magnitude of the induced emf in coil B.

Solution

From Equation 32-2, $\mathcal{E}_B = -M(dI_A/dt) = -M(2bt)$, where the negative value means the emf opposes the change in current.

Problem

59. In the circuit of Fig. 32-8a, take $\mathcal{E}_0 = 5.0$ V and $R = 1.8\ \Omega$. At 2.5 s after the switch is closed, the circuit current is 250 mA. Find the inductance.

Solution

Equation 32-8 for the build-up of current in the RL circuit gives $L = -Rt/\ln(1 - IR/\mathcal{E}_0) = -(1.8\ \Omega)\times(2.5\text{ s})/\ln(1 - 0.25\times1.8/5) = 47.7$ H.

Problem

61. In Fig. 32-13a, take $\mathcal{E}_0 = 25$ V, $R_1 = 1.5\ \Omega$, and $R_2 = 4.2\ \Omega$. What is the voltage across R_2 (a) immediately after the switch is first closed and (b) a long time after the switch is closed? (c) Long after the switch is closed it is again opened. Now what is the voltage across R_2?

Solution

The circuit is analyzed in Example 32-6 at the instants of time mentioned in this problem, so all that is necessary here is to find numerical values. (a) At the moment the switch is first closed, $I = \mathcal{E}_0/(R_1 + R_2) = 25\text{ V}/(1.5 + 4.2)\ \Omega = 4.39$ A, and $V_2 = (4.39\text{ A})(4.2\ \Omega) = 18.4$ V. (b) When the current is steady, the current through R_2 is zero (since the inductor is an ideal one). (c) When the switch is reopened, $V_2 = \mathcal{E}_0R_2/R_1 = (25\text{ V})(4.2/1.5) = 70.0$ V, momentarily.

Problem

63. A wire of radius R carries a current I distributed uniformly over its cross section. Find an expression for the magnetic energy per unit length in the region from R to $100R$.

Solution

The magnetic field from a long straight wire has cylindrical symmetry. A thin coaxial cylindrical shell of length ℓ, radius r, and thickness dr, has volume $dV = 2\pi r\ell\ dr$ and magnetic energy density $u_B = B^2/2\mu_0$. Outside the wire, $B = \mu_0 I/2\pi r$ (see Equation 30-8), so the energy per unit length in a volume extending between $r = R$ and $100R$ is

$$\frac{U}{\ell} = \int \frac{u_B dV}{\ell} = \int_R^{100R} \frac{\mu_0 I^2}{8\pi^2 r^2} 2\pi r\ dr = \frac{\mu_0 I^2}{4\pi} \ln 100.$$

Supplementary Problems

Problem

65. (a) Use the result of Problem 55 to determine the inductance of a toroid. (b) Show that your result reduces to the inductance of a long solenoid when $R \gg \ell$.

Solution

(a) Since $U = \frac{1}{2}LI^2$, dividing the expression for U in Problem 55 by $\frac{1}{2}I^2$, we find $L = (\mu_0 N^2/2\pi)\ell\times\ln(1 + \ell/R)$. (b) For $\ell \ll R, \ln(1 + \ell/R) \approx \ell/R$, so $L \approx \mu_0 N^2\ell^2/2\pi R$. Since $\ell^2 = A$ is the cross-sectional area, $2\pi R = \ell_{\text{Solenoid}}$ is the length, and $n = N/\ell_{\text{Solenoid}}$ is the number of turns per unit length of the toroidal solenoid, this result is approximately the same as Equation 32-4.

Problem

67. (a) Use Equation 32-9 to write an expression for the power dissipation in the resistor as a function of time, and (b) integrate from $t = 0$ to $t = \infty$ to show that the total energy dissipated is equal to the energy initially stored in the inductor, namely, $\frac{1}{2}LI_0^2$.

Solution

Equation 32-9 gives the current decaying through a resistor connected to an inductor carrying an initial current I_0. (a) The instantaneous power dissipated in the resistor is $\mathcal{P}_R = I^2R = I_0^2\ Re^{-2Rt/L}$. (b) In a time interval dt, the energy dissipated is $dU = \mathcal{P}_R\ dt$, so the total energy dissipated is

$$U = \int_0^\infty I_0^2 Re^{-2Rt/L} dt = I_0^2 R \left. \frac{e^{-2Rt/L}}{(-2R/L)} \right|_0^\infty$$
$$= \frac{I_0^2 RL}{2R} = \frac{1}{2}LI_0^2.$$

This is precisely the energy initially stored in the inductor.

Problem

69. An electric field and a magnetic field have the same energy density. Obtain an expression for the ratio E/B, and evaluate this ratio numerically. What are its units? Is your answer close to any of the fundamental constants listed inside the front cover?

Solution

The combination of Equations 26-3 and 32-11 implies that if $u_E = \frac{1}{2}\varepsilon_0 E^2 = u_B = B^2/2\mu_0$, then $E/B = 1/\sqrt{\mu_0\varepsilon_0}$. Numerically, $\mu_0 = 4\pi\times10^{-7}\text{ N/A}^2$ and $(1/4\pi\varepsilon_0) \approx 9\times10^9\text{ N·m}^2/\text{ C}^2$, so $1/\sqrt{\mu_0\varepsilon_0} \approx \sqrt{(9\times10^9\text{ N·m}^2/\text{C}^2)/(10^{-7}\text{ N/A}^2)} = 3\times10^8$ m/s, which is, in fact, the speed of light (see Section 34-5).

Problem

71. The switch in the circuit of Fig. 32-27 is closed at time $t = 0$, at which instant the inductor current is zero. Write the loop and node laws for this circuit, and show that they are satisfied if the inductor current is given by $I = (\mathcal{E}_0/R_1)(1 - e^{-R_\parallel t/L})$, where $R_\parallel$ is the resistance of R_1 and R_2 were they connected in parallel.

Solution

The node law (applied to either node, with currents as shown superimposed on Fig. 32-27) gives $I_1 - I_2 - I_L = 0$. Two independent loop equations (for both loops containing $\mathcal{E}_0$) are $\mathcal{E}_0 - I_1R_1 - I_2R_2 = 0$ and $\mathcal{E}_0 - I_1R_1 + \mathcal{E}_L = 0$. Here, $\mathcal{E}_L = -L(dI_L/dt)$ is the inductor's induced emf. Note that I_1 and I_2 can be determined from I_L, since $I_1 = (\mathcal{E}_0 + \mathcal{E}_L)/R_1$ and $I_2 = I_1 - I_L$. These currents automatically satisfy the node equation and the second loop equation, so we need only check that the expression given for I_L satisfies the first loop equation. The equation can be written entirely in terms of I_L: $0 = \mathcal{E}_0 - I_1R_1 - I_2R_2 = \mathcal{E}_0 - (\mathcal{E}_0 + \mathcal{E}_L) - (I_1 - I_L)R_2 = -\mathcal{E}_L - (\mathcal{E}_0 + \mathcal{E}_L)(R_2/R_1) + I_LR_2 = -\mathcal{E}_L(1 + R_2/R_1) + I_LR_2 - \mathcal{E}_0R_2/R_1$. If we divide by R_2, substitute for $\mathcal{E}_L$, and use $R_\parallel = R_1R_2/(R_1 + R_2)$, the equation becomes $(L/R_\parallel)(dI_L/dt) + I_L - \mathcal{E}_0/R_1 = 0$. Substituting the given expression for I_L, we, indeed, find that

$$\frac{L}{R_\parallel}\left(-\frac{\mathcal{E}_0}{R_1}\right)\left(-\frac{R_\parallel}{L}e^{-R_\parallel t/L}\right) + \frac{\mathcal{E}_0}{R_1}(1 - e^{-R_\parallel t/L}) - \frac{\mathcal{E}_0}{R_1} = 0.$$

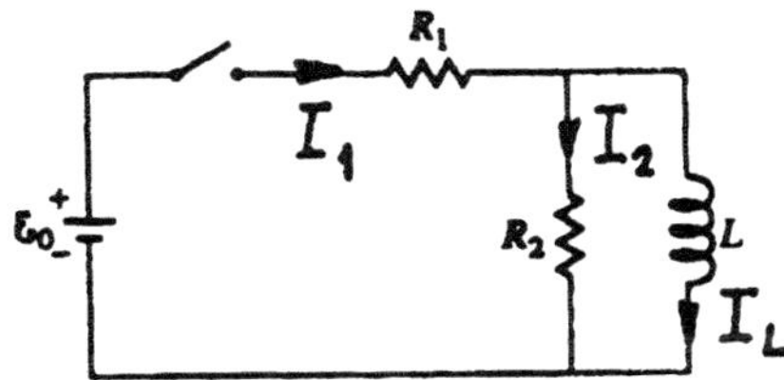

FIGURE 32-27 Problem 71 Solution.

CHAPTER 33 ALTERNATING-CURRENT CIRCUITS

ActivPhysics can help with these problems: Activities 14.2, 14.3

Section 33-1: Alternating Current

Problem

1. Much of Europe uses AC power at 230 V rms and 50 Hz. Express this AC voltage in the form of Equation 33-3, taking $\phi = 0$.

Solution

Use of Equations 33-1 and 2 allows us to write $V_p = \sqrt{2}\, V_{\text{rms}} = \sqrt{2}(230\text{ V}) = 325$ V, and $\omega = 2\pi f = 2\pi(50\text{ Hz}) = 314\text{ s}^{-1}$. Then the voltage expressed in the form of Equation 33-3 is $V(t) = (325\text{ V})\times \sin[(314\text{ s}^{-1})t]$.

Problem

3. An oscilloscope displays a sinusoidal signal whose peak-to-peak voltage (see Fig. 33-1) is 28 V. What is the rms voltage?

Solution

As shown in Fig. 33-1, the peak-to-peak voltage is twice the peak voltage, so Equation 33-1 gives $V_{\text{rms}} = V_p/\sqrt{2} = V_{p\text{-}p}/2\sqrt{2} = 28\text{ V}/2\sqrt{2} = 9.90$ V.

Problem

5. An AC current is given by $I = 495 \sin(9.43t)$, with I in milliamperes and t in milliseconds. Find (a) the rms current and (b) the frequency in Hz.

Solution

Comparison of the current with Equation 33-3 shows that its amplitude and angular frequency are $I_p = 495$ mA and $\omega = 9.43\text{ (ms)}^{-1}$. Application of Equations 33-1 and 2 give (a) $I_{\text{rms}} = 495\text{ mA}/\sqrt{2} = 350$ mA, and (b) $f = 9.43/2\pi(\text{ms}) = 1.50$ kHz.

Problem

7. The rms amplitude is defined as the square root of the average of the square of the signal. For a periodic function, the time average is the integral over one period, divided by the period. For a sinusoidal voltage given by $V = V_p \sin\omega t$, show explicitly that $V_{\text{rms}} = V_p/\sqrt{2}$.

Solution

The period of a sinusoidal signal is $T = 2\pi/\omega$ (Equation 15-6), so

$$\begin{aligned} V_{\text{rms}}^2 = \langle V^2 \rangle &= \frac{1}{T}\int_0^T (V_p^2 \sin^2\omega t)\, dt \\ &= \frac{V_p^2}{2T}\int_0^T (1 - \cos 2\omega t)\, dt \\ &= \frac{V_p^2}{2T}\left[T - \left|\frac{1}{2\omega}\sin 2\omega t\right|_0^T\right] = \frac{V_p^2}{2}. \end{aligned}$$

(We used an identity from Appendix A.) A graphical argument that $\langle \sin^2\omega t\rangle = \frac{1}{2}$ is suggested in Fig. 16-16.

Problem

9. How are the rms and peak voltages related for the triangle wave in Fig. 33-29? See Problem 7.

Solution

Define the zero of time to coincide with a positive apex of the waveform, which has period T, as drawn upon Fig. 33-29. The analytic form of the voltage signal (for $0 \le t \le T$) is

$$V(t) = V_p \begin{cases} -(4t/T) + 1, & 0 \le t \le \frac{1}{2}T \\ (4t/T) - 3, & \frac{1}{2}T \le t \le T. \end{cases}$$

Because the negative part of the waveform is the reflection of the positive part, the square of the waveform has period $\frac{1}{2}T$. Thus, the average over T equals the average over $\frac{1}{2}T$, or

$$\begin{aligned} \langle V^2 \rangle &= \frac{1}{T}\int_0^T V^2 dt = \frac{1}{\frac{1}{2}T}\int_0^{T/2} V^2 dt \\ &= \frac{2V_p^2}{T}\int_0^{T/2}\left(\frac{16t^2}{T^2} - \frac{8t}{T} + 1\right) dt \\ &= \frac{2V_p^2}{T}\left[\frac{16}{3T^2}\left(\frac{T}{2}\right)^3 - \frac{8}{2T}\left(\frac{T}{2}\right)^2 + \frac{T}{2}\right] \\ &= 2V_p^2\left(\frac{2}{3} - 1 + \frac{1}{2}\right) = \frac{V_p^2}{3}. \end{aligned}$$

Consequently, $V_{\text{rms}} = \sqrt{\langle V^2\rangle} = V_p/\sqrt{3}$. (We used a symmetry of the waveform to reduce the number of integrations by half.)

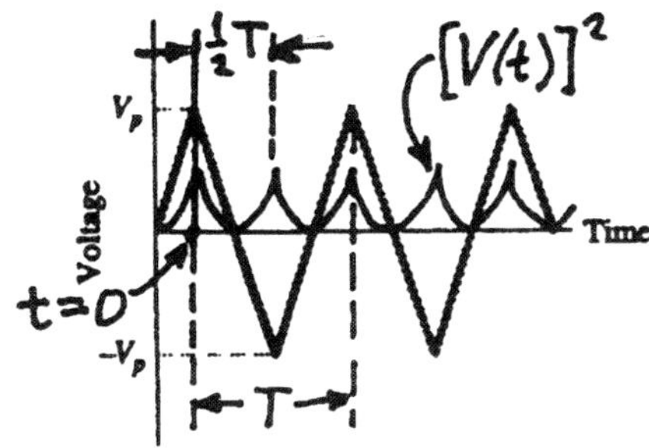

FIGURE 33-29 Problem 9 Solution.

Problem

11. The most general expression for a sinusoidal AC current may be written either $I = I_1 \sin\omega t + I_2 \cos\omega t$ or $I = I_p \sin(\omega t + \phi)$. Find relations between I_1, I_2, I_p, and ϕ that make these expressions equivalent. (See Appendix A for trig identities.)

Solution

Write $I_p \sin(\omega t + \phi) = I_p(\sin\omega t \cos\phi + \cos\omega t \sin\phi) = I_1 \sin\omega t + I_2 \cos\omega t$. The sine and cosine functions are independent, so their coefficients are equal (or, since the equality holds for all t, consider $\omega t = 0$ and $\omega t = \pi/2$.) Thus, $I_1 = I_p \cos\phi$ and $I_2 = I_p \sin\phi$. Inverting these relations, $I_p = \sqrt{I_1^2 + I_2^2}$ and $\phi = \tan^{-1}(I_2/I_1)$, where one must consider the signs of I_1 and I_2 to decide in which quadrant ϕ belongs.

Section 33-2: Circuit Elements in AC Circuits

Problem

13. What is the rms current in a 1.0-μF capacitor connected across the 120-V rms, 60-Hz AC line?

Solution

Equation 33-5 can be used with the rms current and voltage, since both are $1/\sqrt{2}$ times their peak values. Thus, $I_{rms} = \omega C V_{rms} = (2\pi \times 60\text{ Hz})(1\ \mu\text{F})(120\text{ V}) = 45.2\text{ mA}$.

Problem

15. Find the reactance of a 3.3-μF capacitor at (a) 60 Hz, (b) 1.0 kHz, and (c) 20 kHz.

Solution

Equation 33-5 gives $X_C = 1/\omega C = 1/2\pi f(3.3\ \mu\text{F}) = 48.2\text{ k}\Omega\cdot\text{Hz}/f$. For the given frequencies, $X_C =$ (a) 804 Ω, (b) 48.2 Ω, (c) 2.41 Ω respectively. (One can see that a capacitor has the greatest effect at low frequency.)

Problem

17. A capacitor and a 1.8-kΩ resistor pass the same current when each is separately connected across a 60-Hz power line. What is the capacitance?

Solution

The currents (rms or peak) are the same if $X_C = R = 1/\omega C$, so $C = 1/\omega R = [2\pi(60\text{ Hz})(1.8\text{ k}\Omega)]^{-1} = 1.47\ \mu\text{F}$.

Problem

19. A 50-mH inductor is connected across a 10-V rms AC generator, and an rms current of 2.0 mA flows. What is the generator frequency?

Solution

From Equation 33-7, $f = \omega/2\pi = V_p/2\pi I_p L$. Since the ratio of the peak values of voltage and current is the same as that of the rms values, $f = 10\text{ V}/2\pi(2\text{ mA})\times(50\text{ mH}) = 15.9\text{ kHz}$.

Problem

21. A 1.2-μF capacitor is connected across a generator whose output is given by $V = V_p \sin 2\pi f t$, where $V_p = 22$ V, $f = 60$ Hz, and t is in seconds. (a) What is the peak current? (b) What are the magnitudes of the voltage and (c) the current at $t = 6.5$ ms?

Solution

(a) $I_p = V_p\omega C = (22\text{ V})(377\text{ s}^{-1})(1.2\ \mu\text{F}) = 9.95\text{ mA}$ (see Equation 33-5 and Example 33-1). (b) $V(6.5\text{ ms}) = V_p \sin\omega t = (22\text{ V})\sin[(377\text{ s}^{-1})(6.5\text{ ms})] = 14.0\text{ V}$. (Remember that ωt is in radians.) (c) In a capacitor, the current leads the voltage by 90°, so $I(6.5\text{ ms}) = I_p \sin(\omega t + \pi/2) = (9.95\text{ mA})\cos[(377\text{ s}^{-1})(6.5\text{ ms})] = -7.67\text{ mA}$ (the magnitude is 7.67 mA).

Problem

23. What is the maximum charge on the plates of a 16-μF capacitor connected across the 120-V rms, 60-Hz AC power line?

Solution

The charge on the capacitor is $q(t) = CV_C(t)$, so the maximum charge (for sinusoidal voltage) is $q_{max} = CV_{C,p} = \sqrt{2}\, CV_{C,rms} = \sqrt{2}\,(16\ \mu\text{F})(120\text{ V}) = 2.72\text{ mC}$ (independent of frequency).

Problem

25. A 0.75-H inductor is in series with a flourescent lamp, and the series combination is across the

120-V rms, 60-Hz power line. If the rms inductor voltage is 90 V, what is the rms lamp current?

Solution

In a series circuit, the same current flows through the inductor and lamp. Therefore, since the ratio of the rms quantities for a given circuit element equals that of the peak values, Equation 33-7 gives $I_{\rm rms} = V_{\rm rms}/\omega L = 90\ \text{V}/(2\pi \times 60\ \text{Hz})(0.75\ \text{H}) = 318\ \text{mA}$.

Section 33-3: *LC* Circuits

Problem

27. Find the resonant frequency of an LC circuit consisting of a 0.22-μF capacitor and a 1.7-mH inductor.

Solution

Equations 33-2 and 11 give $f = 1/2\pi\sqrt{LC} = 1/2\pi\sqrt{(0.22\ \mu\text{F})(1.7\ \text{mH})} = 8.23\ \text{kHz}$.

Problem

29. You have a 2.0-mH inductor and wish to make an LC circuit whose resonant frequency spans the AM radio band (550 kHz to 1600 kHz). What range of capacitance should your variable capacitor cover?

Solution

The resonant frequency of an LC circuit is (Equation 33-11) $\omega = 1/\sqrt{LC}$, so the capacitance should cover a range from $C = 1/\omega^2 L = 1/(2\pi \times 550\ \text{kHz})^2(2\ \text{mH}) = 41.9\ \text{pF}$ down to $C = 1/(2\pi \times 1.6\ \text{MHz})^2(2\ \text{mH}) = 4.95\ \text{pF}$.

Problem

31. You want to use an LC circuit in a timing application. The circuit is to start with the capacitor fully charged, and the voltage should drop to zero in 15 s. You have available a 25-H inductor. What capacitance should you use?

Solution

The capacitor voltage drops to zero in one quarter of a period, so $\omega = 2\pi/T = 2\pi/(4 \times 15\ \text{s})$. From Equation 33-11, $C = 1/\omega^2 L = (60\ \text{s}/2\pi)^2/(25\ \text{H}) = 3.65\ \text{F}$.

Problem

33. An LC circuit includes a 0.025-μF capacitor and a 340-μH inductor. (a) If the peak voltage on the capacitor is 190 V, what is the peak current in the inductor? (b) How long after the voltage peak does the current peak occur?

Solution

(a) In an LC circuit, the peak current and voltage are related by $I_p = \omega q_p = \omega C V_p = C V_p/\sqrt{LC} = V_p\sqrt{C/L}$ (see Example 33-3). Thus, $I_p = (190\ \text{V}) \times \sqrt{0.025\ \mu\text{F}/340\ \mu\text{H}} = 1.63\ \text{A}$. (b) The current peaks one quarter of a period after the voltage, or $\Delta t = \frac{1}{4}T = \frac{1}{4}(2\pi/\omega) = \frac{1}{2}\pi\sqrt{LC} = (\pi/2)\sqrt{0.025\ \mu\text{F} \times 340\ \mu\text{H}} = 4.58\ \mu\text{s}$.

Problem

35. At the instant when the electric and magnetic energies are equal in the LC circuit of Problem 33, the current is 540 mA. (a) What is the instantaneous voltage? Find (b) the peak voltage, (c) the peak current, and (d) the total energy.

Solution

(a) The energies are instantaneously equal when $\frac{1}{2}CV(t)^2 = \frac{1}{2}LI(t)^2$, or $V(t) = I(t)\sqrt{L/C} = (540\ \text{mA}) \times \sqrt{340\ \mu\text{H}/0.025\ \mu\text{F}} = 63.0\ \text{V}$. (b) The times when the energies are equal are given by $\sin^2\omega t = \cos^2\omega t$, so $|\sin\omega t| = |\cos\omega t| = 1/\sqrt{2}$ at those times (recall that $\sin^2 + \cos^2 = 1$). From $V(t) = V_p\cos\omega t$, we get $V_p = (63.0\ \text{V})\sqrt{2} = 89.1\ \text{V}$. (c) Similarly, $I_p = (540\ \text{mA})\sqrt{2} = 764\ \text{mA}$. (d) The total energy is $\frac{1}{2}CV_p^2 = \frac{1}{2}LI_p^2 = \frac{1}{2}CV(t)^2 + \frac{1}{2}LI(t)^2 = LI(t)^2 = (340\ \mu\text{H})(540\ \text{mA})^2 = 99.1\ \mu\text{J}$. (We chose to evaluate the total energy from twice the inductor's energy, at the instant the capacitor and inductor energies are equal, because these numerical values were given.)

Problem

37. One-eighth of a cycle after the capacitor in an LC circuit is fully charged, what are each of the following as fractions of their peak values: (a) capacitor charge, (b) energy in the capacitor, (c) inductor current, (d) energy in the inductor?

Solution

The equations in Section 33-3 give the desired quantities, which we evaluate when $\omega t = \omega(T/8) = 2\pi/8 = \frac{1}{4}\pi = 45°$ (i.e., $\frac{1}{8}$ cycle). (Note that phase constant zero corresponds to a fully charged capacitor at $t = 0$.) (a) From Equation 33-10, $q/q_p = \cos 45° = 1/\sqrt{2}$. (b) From the equation for electric energy, $U_E/U_{E,p} = \cos^2 45° = 1/2$. (c) From Equation 33-12, $I/I_p = -\sin 45° = -1/\sqrt{2}$. (The direction of the current is away from the positive capacitor plate at $t = 0$.) (d) From the equation for magnetic energy, $U_B/U_{B,p} = \sin^2 45° = 1/2$.

Problem

39. The 2000-μF capacitor in Fig. 33-30 is initially charged to 200 V. (a) Describe how you would manipulate switches A and B to transfer all the energy from the 2000-μF capacitor to the 500-μF capacitor. Include the times you would throw the switches. (b) What will be the voltage across the 500-μF capacitor once you've finished?

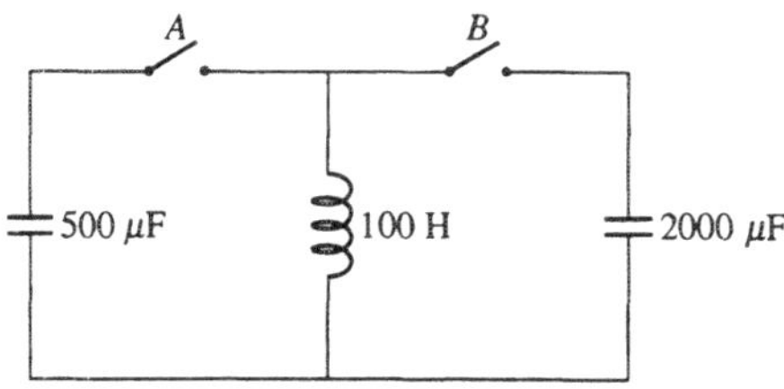

FIGURE 33-30 Problem 39.

Solution

(a) The energy initially stored in the first capacitor is $\frac{1}{2}(2\text{ mF})(200\text{ V})^2 = 40$ J. First close switch B for one quarter of a period of the LC circuit containing the 2000 μF capacitor, or $t_B = \frac{1}{4}T_B = \frac{1}{4}(2\pi/\omega_B) = \frac{1}{2}\pi\sqrt{LC_B} = \frac{1}{2}\pi\sqrt{(100\text{ H})(2\text{ mF})} = 702$ ms. This transfers 40 J to the inductor. Then open switch B and close switch A for one quarter of a period of the LC circuit containing the 500 μF capacitor, or $t_A = \frac{1}{2}\pi\sqrt{(100\text{ H})(0.5\text{ mF})} = \frac{1}{2}t_B = 351$ ms. This transfers 40 J to the second capacitor from the inductor. Finally, open switch A. (b) When the second capacitor has 40 J of stored energy, its voltage is $\sqrt{2(40\text{ J})/(0.5\text{ mF})} = 400$ V.

Problem

41. A damped RLC circuit includes a 5.0-Ω resistor and a 100-mH inductor. If half the initial energy is lost after 15 cycles, what is the capacitance?

Solution

If only half the energy is lost after 15 cycles, the damping is small and the energy varies like the square of Equation 33-13, namely $U_{\text{tot}} = U_p e^{-Rt/L}\cos^2\omega t$. (The energy time constant is L/R, one half the charge time constant.) After 15 cycles, $t = 15\ T = 15(2\pi/\omega)$, the fraction of energy remaining is $\frac{1}{2} = e^{-15RT/L}\times \cos^2 30\pi = e^{-15RT/L}$. Take logarithms and use $\omega = 1/\sqrt{LC}$ to get $L\ln 2 = 15RT = 30\pi R\sqrt{LC}$, from which we find

$$C = \left(\frac{\ln 2}{30\pi R}\right)^2 L = \left(\frac{\ln 2}{30\pi \times 5\ \Omega}\right)^2 (100\text{ mH}) = 0.216\ \mu\text{F}.$$

Section 33-4: Driven RLC Circuits and Resonance

Problem

43. If the speaker system of Example 33-4 is driven by a 10-V peak, 1.0-kHz sine wave, what will be the peak voltage across the capacitor?

Solution

The peak capacitor voltage in a series RLC circuit at resonance ($\omega_0 = 1$ kHz in Example 33-4) is $V_{C,p} = I_pX_C = V_p/\omega_0RC = 10\text{ V}/[2\pi(1\text{ kHz})(8.51\ \Omega)\times (11.5\ \mu\text{F})] = 16.2$ V. (Alternatively, $V_{C,p} = V_p\omega_0L/R = (10\text{ V})2\pi(1\text{ kH})(2.2\text{ mH})/(8.51\ \Omega)$, since $X_C = X_L$ at resonance. Finally, since $\omega_0 = 1/\sqrt{LC}$, $V_{C,p} = (V_p/R)\sqrt{L/C} = (10\text{ V}/8.51\ \Omega)\sqrt{2.2\text{ mH}/11.5\ \mu\text{F}}$. We calculated a value of $R = 8.51\ \Omega$ without the round-off errors in Example 33-4.)

Problem

45. TV channel 2 occupies the frequency range from 54 MHz to 60 MHz. A series RLC tuning circuit in a TV receiver includes an 18-pF capacitor and resonates in the middle of the channel 2 band. (a) What is the inductance? (b) To let the whole signal in, the resonance curve must be broad enough that the current throughout the band be no less than 70% of the current at the resonant frequency. What constraint does this place on the circuit resistance?

Solution

(a) The condition for resonance in a series RLC circuit requires $L = 1/\omega_0^2C = 1/(2\pi\times 57\text{ MHz})^2(18\text{ pF}) = 0.433\ \mu$H. (b) The peak current at any frequency is $I_p = V_p/Z$, while at resonance, $I_{\text{res}} = V_p/R$. Thus, $I_p/I_{\text{res}} = R/Z = 1/\sqrt{1+(X_L-X_C)^2/R^2}$ (see Equation 33-14). If this ratio is required to be not less than 70%, then

$$\left(\frac{X_L - X_C}{R}\right)^2 \le \left(\frac{1}{0.7}\right)^2 - 1,\ \text{or}\ R \ge \sqrt{\frac{49}{51}}\,|X_L - X_C|.$$

The reactance can be expressed in several equivalent forms, one being in terms of the given data and the resonant frequency:

$$|X_L - X_C| = \left|\omega L - \frac{1}{\omega C}\right| = \sqrt{\frac{L}{C}}\left|\omega\sqrt{LC} - \frac{1}{\omega\sqrt{LC}}\right| = \frac{1}{\omega_0 C}\left|\frac{\omega}{\omega_0} - \frac{\omega_0}{\omega}\right|.$$

If this is evaluated at the lower band edge (54 MHz), one gets

$$R \geq \frac{\sqrt{49/51}}{(2\pi \times 57 \text{ MHz})(18 \text{ pF})}\left(\frac{57}{54} - \frac{54}{57}\right) = 16.4 \ \Omega.$$

(The upper band edge, 60 MHz, gives a weaker limit, $R \geq 15.6 \ \Omega$.)

Problem

47. A 2.0-H inductor and a 3.5-μF capacitor are connected in series with a 50-Ω resistor, and the combination is connected to an AC generator supplying 24 V peak at 60 Hz. (a) At the instant the generator voltage is at its peak, what is the instantaneous voltage across each circuit element? Show explicitly that these sum to the generator voltage. (b) If rms voltmeters are connected across each of the three components, what will they read? Do their readings sum to the rms generator voltage? Does this contradict the loop law?

Solution

Before calculating numerical results, we display the general time-dependent voltages in an RLC series circuit, which can be derived from the phasor diagram in Fig. 33-17*b* (the angle between V_p and the horizontal axis is ωt). Note that the phase constants in Equation 33-3 (for the applied voltage and current, i.e., the resistor voltage) are related to the phase difference in Equation 33-16 by $\phi_V = 0$ and $\phi_I = -\phi$. (See paragraphs following Equation 33-16.)

$$V = V_p \sin\omega t \qquad V_{R,p} = I_p R$$

$$Z = \sqrt{R^2 + (X_L - X_C)^2}$$

$$V_R = V_{R,p}\sin(\omega t - \phi) \qquad V_{C,p} = I_p X_C$$

$$\phi = \tan^{-1}\left(\frac{X_L - X_C}{R}\right)$$

$$V_C = V_{C,p}\sin(\omega t - \phi - \tfrac{1}{2}\pi) \quad V_{L,p} = I_p X_L$$

$$-\tfrac{1}{2}\pi \leq \phi \leq \tfrac{1}{2}\pi$$

$$V_L = V_{L,p}\sin(\omega t - \phi + \tfrac{1}{2}\pi) \qquad I_p = V_p/Z$$

(a) At the instant when $V = V_p = 24$ V, $\omega t = 90°$. From the given values of R, L, C, ω, and V_p, we can calculate ϕ and the instantaneous voltages: $V_R = (23.93 \text{ V})\sin(90° + 4.458°) = 23.86$ V; $V_C = (362.7 \text{ V})\sin 4.458° = 28.19$ V; $V_L = (360.8 \text{ V})\times \sin 184.458° = -28.05$ V. Four significant figures are necessary to verify the loop law, $V_R + V_C + V_L = 24.00$ V, in this case. (b) The rms voltages are $1/\sqrt{2}$ times the peak values found in part (a) (approximately 16.9, 256, and 255 volts for R, C, and L respectively). Of course, since the instantaneous voltages have different phases, the sum of the rms voltages does not equal $V_{\text{rms}} = V_p/\sqrt{2} = 17.0$ V.

Problem

49. Figure 33-31 shows the phasor diagram for an RLC circuit. (a) Is the driving frequency above or below resonance? (b) Complete the diagram by adding the applied voltage phasor, and from your diagram determine the phase difference between applied voltage and current.

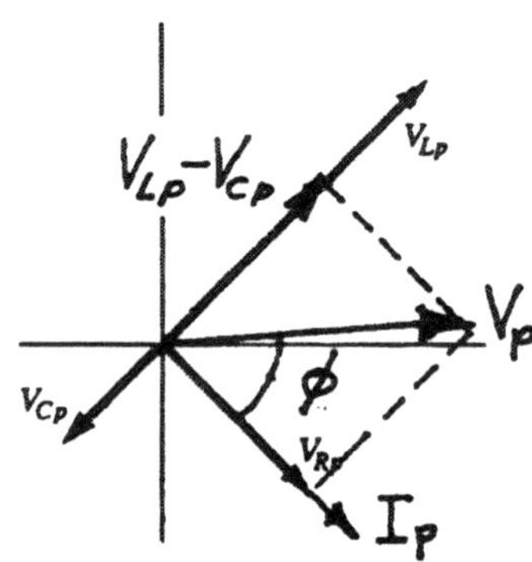

FIGURE 33-31 Problem 49 Solution.

Solution

(a) From the observation that $V_{L,p} = I_p\omega L > V_{C,p} = I_p/\omega C$, we conclude that the frequency is above resonance, $\omega^2 > 1/LC$. (b) The applied voltage phasor is the vector sum of the resistor, capacitor, and inductor voltage phasors, as drawn on Fig. 33-31. The current is in phase with the voltage across the resistor, in this case lagging the applied voltage (since $\phi = \tan^{-1}[(V_{L,p} - V_{C,p})/V_{R,p}] > 0$) by approximately 50° (as estimated from Fig. 33-31).

Problem

51. For the circuit of Problem 46, find the phase relation between applied voltage and current at frequencies of (a) 550 Hz and (b) 700 Hz.

Solution

The phase constant (relative phase of the current and the applied voltage) for a series RLC circuit is given by $\phi = \tan^{-1}[(\omega L - 1/\omega C)/R]$ (Equation 33-16). For the given values of R, L, and C in Problem 46, and the lower frequency $\phi = \tan^{-1}[(2\pi(550 \text{ Hz})(50 \text{ mH}) - (2\pi)^{-1}(550 \text{ Hz})^{-1}(1.5 \ \mu\text{F})^{-1})/10 \ \Omega] = -63.6°$. As expected, for a frequency below resonance ($\omega_0 = 2\pi(581 \text{ Hz})$, the current leads the voltage ($X_C > X_L$ so $\phi < 0$). A similar calculation at 700 Hz gives $\phi = +81.7°$, or the current lags the voltage at a frequency above resonance ($X_L > X_C$ so $\phi > 0$).

Problem

53. An electric drill draws 4.6 A rms at 120 V rms. If the current lags the voltage by 25°, what is the drill's power consumption?

Solution

The average power consumed by an AC circuit is given by Equation 33-17, $\mathcal{P}_{av} = V_{rms}I_{rms}\cos\phi = (120\text{ V})\times(4.6\text{ A})\cos(+25°) = 500\text{ W}$.

Problem

55. A series RLC circuit has power factor 0.80 and impedance 100 Ω at 60 Hz. (a) What is the circuit resistance? (b) If the inductance is 0.10 H, what is the resonant frequency?

Solution

(a) The geometry of Fig. 33-17*b*, with the peak voltages expressed in terms of the peak current, shows that $\cos\phi = V_{R,p}/V_p = (I_pR)/(I_pZ) = R/Z$. (Alternatively, use Equations 33-15 and 16 and the trigonometric identity $\sec^2\phi = 1 + \tan^2\phi$.) Therefore, $R = Z\cos\phi = (100\ \Omega)(0.8) = 80\ \Omega$. (b) The reactance of the circuit can be expressed in terms of the inductance, the resonant frequency, and the given values by the use of Equation 33-15:

$$\begin{aligned} X_L - X_C &= \pm\sqrt{Z^2 - R^2} = \pm\sqrt{100^2 - 80^2}\ \Omega = \pm 60\ \Omega \\ &= \omega L - 1/\omega C = \omega L(1 - 1/\omega^2 LC) \\ &= \omega L(1 - \omega_0^2/\omega^2),\ \text{or}\ \omega_0^2/\omega^2 = 1 \pm (60\ \Omega/\omega L). \end{aligned}$$

Since $\omega L = (2\pi \times 60\text{ Hz})(0.1\text{ H}) = 37.7\ \Omega < 60\ \Omega$, we can discard the unphysical solution (with $\omega_0^2/\omega^2 < 0$) to find $\omega_0 = \omega\sqrt{1 + (60/37.7)}$, or $f_0 = 60\text{ Hz}\sqrt{2.59} = 96.6\text{ Hz}$. (If the value of ωL has been >60 Ω, there would have been two physically acceptable solutions.)

Problem

57. A power plant produces 60-Hz power at 365 kV rms and 200 A rms. The plant is connected to a small city by a transmission line with total resistance 100 Ω. What fraction of the power is lost in transmission if the city's power factor is (a) 1.0 or (b) 0.60? (c) Is it more economical for the power company if the load has a large power factor or a small one? Explain.

Solution

(a) We assume that the average power supplied to the city is $\mathcal{P}_{av} = I_{rms}V_{rms}\cos\phi = (200\text{ A})(365\text{ kV})(1) = 73.0\text{ MW}$, and that the average power lost in the transmission line is $\Delta\mathcal{P} = I_{rms}^2 R = (200\text{ A})^2(100\ \Omega) = 4\text{ MW}$. Thus, the percent lost is $(4/73)\times 100 \simeq 5.5\%$.
(b) If the power factor in part (a) were 0.6 instead of 1.0, the percent lost would be $5.5\%/0.6 = 9.1\%$.
(c) Since $\Delta\mathcal{P}$ is a constant, the larger $\mathcal{P}_{av}$ (which is proportional to $\cos\phi$), the smaller the fraction of power lost, $\Delta\mathcal{P}/\mathcal{P}_{av}$. A large power factor is better for the power plant owners.

Section 33-6: Transformers and Power Supplies

Problem

59. A transformer steps up the 120-V rms AC power line voltage to 23 kV rms for a TV picture tube. If the rms current in the primary is 1.0 A, and the transformer is 95% efficient, what is the secondary current?

Solution

Only 95% of the power in the primary is transformed to power in the secondary, so Equation 33-19 can be modified to yield $(0.95)(120\text{ V})(1\text{ A}) = (23\text{ kV})I_{sec}$, or $I_{sec} = 4.96\ \text{mA (rms)}$.

Problem

61. The transformer in the power supply of Fig. 33-26*a* has an output voltage of 6.3 V rms at 60 Hz, and the capacitance is 1200 μF. (a) With an infinite load resistance, what would be the output voltage of the power supply? (b) What is the minimum load resistance for which the output would not drop more than 1% from this value? Assume that the discharge time in Fig. 33-26*b* is essentially a full cycle.

Solution

(a) With no load present, the capacitor cannot discharge through the transformer because the diode blocks such a current. Therefore, it remains charged to the peak AC voltage, $V_p = \sqrt{2}\,V_{rms} = \sqrt{2}(6.3\text{ V}) = 8.91\text{ V}$. (b) With the load resistance present, the capacitor voltage must not decay to less than 99% of its maximum value over one period of AC ($T = \frac{1}{60}$ s) (see Fig. 33-26*b*). Therefore, $e^{-T/RC} \geq 0.99$, or $R \geq -T/C\ln(0.99) = [-(60\text{ Hz})(1200\ \mu\text{F})\ln(0.99)]^{-1} = 1.38\text{ k}\Omega$. (Note: The voltage for a discharging RC circuit is given by Equation 28-8.)

Paired Problems

Problem

63. A sine-wave generator delivers a signal whose peak voltage is independent of frequency. Two identical capacitors are connected in parallel across the generator, and the generator supplies a peak current I_p at frequency f_1. The capacitors are

then connected in series across the generator. To what frequency should the generator be tuned to bring the current back to I_p?

Solution

For constant $V_p = I_p X_C$ (i.e., independent of frequency), the same peak current will be supplied if the capacitive reactances for the two connections are equal, i.e., $2\pi f_1 C_1 = 2\pi f_2 C_2$. Thus, for parallel and series combinations of two equal capacitors, $f_2 = f_1(C_{\text{parallel}}/C_{\text{series}}) = f_1(C_a + C_b)^2/C_a C_b = 4f_1$. (We left the capacitors general in the intermediate step for application to the next problem.)

Problem

65. The peak current in an oscillating LC circuit is 850 mA. If $L = 1.2$ mH and $C = 5.0$ μF, what is the peak voltage?

Solution

In an LC circuit, the peak electric and magnetic energies are equal (see the analysis in the text), so $\frac{1}{2}CV_p^2 = \frac{1}{2}LI_p^2$. Thus, $V_p = \sqrt{L/C}I_p = \sqrt{1.2 \text{ mH}/5.0 \ \mu\text{F}}(850 \text{ mA}) = 13.2$ V.

Problem

67. An RLC circuit includes a 3.3-μF capacitor and a 27-mH inductor. The capacitor is charged to 35 V, and the circuit begins oscillating. Ten full cycles later the capacitor voltage peaks at 28 V. What is the resistance?

Solution

For the damped oscillations of an RLC circuit, the voltage decays according to Equation 33-13, $V(t) = q(t)/C = V_p e^{-Rt/2L} \cos \omega t$, with frequency given by Equation 33-11. If in ten cycles ($t = 10\ T = 10(2\pi/\omega) = 20\pi\sqrt{LC}$) the peak voltage has decayed from 35 V to 28 V, then $\ln(35/28) = Rt/2L = 10\pi R\sqrt{C/L}$, or $R = (10\pi)^{-1}\sqrt{27 \text{ mH}/3.3 \ \mu\text{F}} \times \ln(35/28) = 642$ mΩ.

Problem

69. A series RLC circuit with $R = 5.5\ \Omega$, $L = 180$ mH, and $C = 0.12$ μF is connected across a sine-wave generator. If the inductor can handle a maximum current of 1.5 A, what is the maximum safe value for the generator's peak output voltage when it is tuned to resonance?

Solution

At resonance, the impedance of a series RLC circuit is $Z = R$, so $V_p = ZI_p = RI_p = (5.5\ \Omega)(1.5 \text{ A}) = 8.25$ V at the maximum safe peak current.

Supplementary Problems

Problem

71. Two capacitors are connected in parallel across a 10-V rms, 10-kHz sine-wave generator, and the generator supplies a total rms current of 30 mA. When the capacitors are rewired in series, the rms generator current drops to 5.5 mA. Find the values of the two capacitances.

Solution

Equation 33-5 gives the rms current when capacitors are connected to an AC generator, $I_{\text{rms}} = V_{\text{rms}}/X_C = \omega C V_{\text{rms}}$. For the parallel connection, 30 mA $= (2\pi \times 10^5 \text{ V/s})(C_1 + C_2)$, while for the series connection, 5.5 mA $= (2\pi \times 10^5 \text{ V/s})C_1C_2/(C_1 + C_2)$. Thus, $C_1 + C_2 = 47.7$ nF and $C_1C_2 = (20.4 \text{ nF})^2$. Eliminate one capacitance from the second equation and substitute into the first equation to obtain a quadratic, $C^2 - (47.7 \text{ nF})C + (20.4 \text{ nF})^2 = 0$, whose two solutions are C_1 and C_2, i.e., $\frac{1}{2}[(47.7 \text{ nF}) \pm \sqrt{(47.7 \text{ nF})^2 - 4(20.4 \text{ nF})^2}] = 11.5$ nF and 36.2 nF. (The equations are symmetric in C_1 and C_2.)

Problem

73. An undriven RLC circuit with inductance L and resistance R starts oscillating with total energy U_0. After N cycles the energy is U_1. Find an expression for the capacitance, assuming the circuit is not heavily damped.

Solution

The total energy in an underdamped oscillating RLC circuit decays as the square of Equation 33-13 ($U = q^2/2C$), or $U_{\text{tot}} = U_0 e^{-Rt/L}$. After N cycles, $t = N(2\pi/\omega) = 2\pi N\sqrt{LC}$ (see Equation 33-11), so $U_1 = U_0 e^{-2\pi RN\sqrt{C/L}}$, and $C = L[\ln(U_0/U_1)/2\pi RN]^2$.

Problem

75. You wish to make a "black box" with two input connections and two output connections, as shown in Fig. 33-33. When you put a 12-V rms, 60-Hz sine wave across the input, a 6.0-V, 60-Hz signal should appear at the output, with the output voltage leading the input voltage by 45°. Design a circuit that could be used in the "black box."

FIGURE 33-33 Problem 75.

Solution

Since the voltage across a resistor leads the voltage across a capacitor in series with it (see Problem 82), and the voltage across an inductor leads the voltage across a resistor in series with it (see Problem 83), either circuit can be adapted to the criteria of the "black box" in this problem. (We include the solution to these three problems below.)

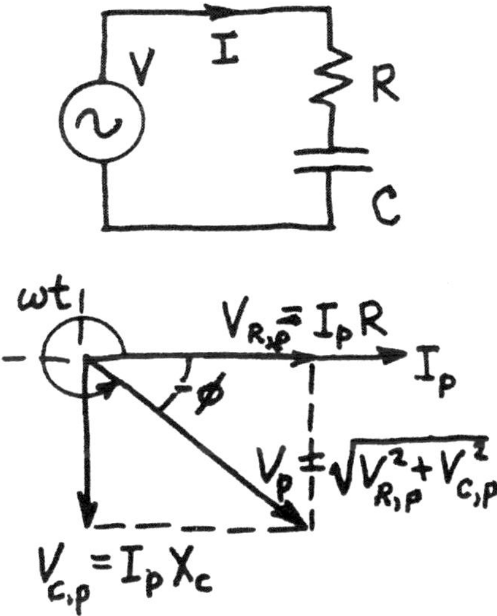

Problem 75 Solution (see Problem 82).

In the circuit diagram for Problem 82, it can be seen that $V = V_R + V_C$ and $I = I_C = I_R$. In the corresponding phasor diagram, V_C lags I by 90°, V_R and I are in phase, and V is the vector sum of these (see Table 33-1). (We drew I horizontally for convenience.) The impedance is

$$Z = \frac{V_p}{I_p} = \frac{\sqrt{V_{R,p}^2 + V_{C,p}^2}}{I_p}$$
$$= \sqrt{R^2 + X_C^2} = \sqrt{R^2 + \frac{1}{\omega^2C^2}},$$

and the phase angle is $\tan\phi = -V_{C,p}/V_{R,p} = -X_C/R = -1/\omega RC$. I always leads V, because $\phi < 0$. (Recall that ϕ is defined by $I = I_p \sin(\omega t - \phi)$ when $V = V_p \sin \omega t$.)

When an inductor in Problem 83 replaces the capacitor of Problem 82, the phasors for V_R and I are still parallel, but V_L leads I by 90°. V is the vector sum of V_R and V_L, so the impedance is

$$Z = \frac{V_p}{I_p} = \frac{\sqrt{V_{R,p}^2 + V_{L,p}^2}}{I_p}$$
$$= \sqrt{R^2 + X_L^2} = \sqrt{R^2 + \omega^2L^2},$$

and the phase angle is $\tan\phi = V_{L,p}/V_{R,p} = X_L/R = \omega L/R$. In this case, I always lags V, because $\phi > 0$. (Negative ϕ is in the same sense as ωt, measured from V.)

The solution to Problem 82 shows that, in a series RC circuit, V_R leads the applied voltage, V, by an angle $\tan^{-1}(1/\omega RC)$, which may be adjusted to 45° if $\omega RC = 1$. The peak voltage across the entire resistance is $V_{R,p} = V_p \cos 45° = V_p/\sqrt{2}$, so if we divide the resistance into two parts, $R_1 + R_2 = R$, with $R_2/R = 1/\sqrt{2}$, then the peak voltage across R_2 will be $(1/\sqrt{2})V_{R,p} = \frac{1}{2}V_p$, as desired (rms voltages have the same ratio as peak voltages).

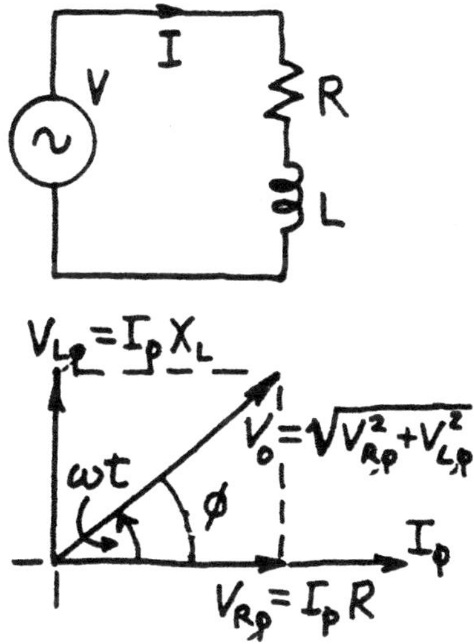

Problem 75 Solution (see Problem 83).

Alternatively, the solution to Problem 83 shows that, in a series RL circuit, V_L leads V by $90° - \tan^{-1}(\omega L/R)$, which equals 45° if $\omega L = R$. Again, $V_{L,p} = V_p/\sqrt{2}$, so if we divide L into $L_1 + L_2$, with $L_2 = L/\sqrt{2}$, the peak voltage across L_2 is $\frac{1}{2}V_p$. Both circuits are sketched below.

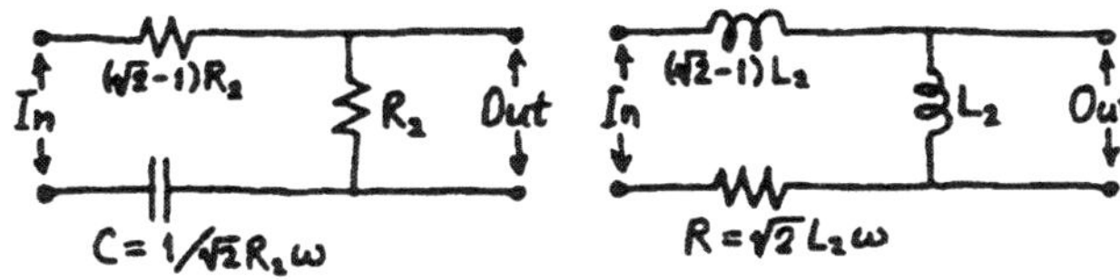

Problem 75 Solution (3).

(Note that V_{out} is the open-circuit output voltage. If a load is connected across the output terminals, the magnitude and the phase of the voltage will be changed accordingly.)

Problem

77. A sine-wave generator with peak output voltage of 20 V is applied across a series RLC circuit. At the resonant frequency of 2.0 kHz the peak current is 50 mA, while at 1.0 kHz it is 15 mA. Find R, L, and C.

Solution

At resonance, $I_p = V_p/R$, so $R = 20\text{ V}/50\text{ mA} = 400\ \Omega$. The impedance at resonance is $Z = R$ [i.e., $X = X_L - X_C = 0$), while at half the resonant frequency $(1\text{ kHz} = \frac{1}{2}(2\text{ kHz})]$, $Z = V_p/I_p = 20\text{ V}/15\text{ mA} = 1.33\text{ k}\Omega = (10/3)R$ (i.e., $|X| = \sqrt{Z^2 - R^2} = \sqrt{(10/3)^2 - 1}\ R = \sqrt{91}\ R/3$.) Therefore,

$$\frac{1}{\omega_0 C} - \omega_0 L = 0, \text{ and } \frac{1}{\frac{1}{2}\omega_0 C} - \frac{1}{2}\omega_0 L = \sqrt{91}\ R/3.$$

These equations can be solved for C and L, with the following result:

$$L = \frac{2\sqrt{91}\ R}{9\omega_0} = \frac{2\sqrt{91}(400\ \Omega)}{9\times 2\pi\times 2\text{ kHz}} = 67.5\text{ mH, and}$$
$$C = (\omega_0^2 L)^{-1} = (2\pi \times 2\text{ kHz})^{-2}(67.5\text{ mH})^{-1} = 93.8\text{ nF}.$$

(Note: Below resonance, $X_C > X_L$.)

Problem

79. A 2.5-H inductor is connected across a 1500-μF capacitor. A 5.0-kg mass is connected to a spring. What should be the spring constant if the mechanical and electrical systems have the same resonant frequency?

Solution

The comparison of Equations 15-4 and 33-9 in the text shows that the resonant frequency of each system is $\omega_0^2 = 1/LC = k/m$ (see Equations 15-12 and 33-11), so $k = m/LC = (5\text{ kg})/(2.5\text{ H})(1.5\text{ mF}) = 1.33\text{ kN/m}$.

Problem

81. For RLC circuits in which the resistance is not too large, the Q factor may be defined as the ratio of the resonant frequency to the difference between the two frequencies where the power dissipated in the circuit is half that dissipated at resonance. Show, using suitable approximations, that this definition leads to the expression $Q = \omega_0 L/R$, with ω_0 the resonant frequency.

Solution

From Equations 33-14 (with rms values), 33-17, and the result in the solution to Problem 55(a), the average power in a series RLC circuit becomes $\langle\mathcal{P}\rangle = I_{rms}V_{rms}\cos\phi = (V_{rms}/Z)V_{rms}(R/Z) = V_{rms}^2 R/Z^2$. From Equation 33-15, one sees that the power falls to half its resonance value (V_{rms}^2/R) when $Z = \sqrt{2}R$, or when $|X_L - X_C| = R$. In terms of the resonant frequency, $\omega_0 = 1/\sqrt{LC}$, this condition becomes

$$\left|\omega L - \frac{1}{\omega C}\right| = L\left|\omega - \frac{\omega_0^2}{\omega}\right| = R, \text{ or } \omega^2 - \omega_0^2 = \pm\frac{R}{L}\omega.$$

The solutions of these quadratics, with $\omega > 0$, are

$$\omega = \frac{1}{2}\left[\pm\frac{R}{L} + \sqrt{\frac{R^2}{L^2} + 4\omega_0^2}\right].$$

If $R/L \ll \omega_0$ (equivalent to $R \ll \sqrt{L/C}$), we can neglect the first term under the square root sign compared to the second, obtaining $\omega \approx \omega_0 \pm R/2L$. The difference between these two values of ω is $\Delta\omega = R/L$, from which the desired expression for Q follows.

Problem

83. Consider a series circuit containing an AC generator, a resistor, and an inductor. Construct a phasor diagram, and derive expressions for the circuit impedance and the phase angle between the applied voltage and the current. Show that the voltage always leads the current.

Solution

See solution to Problem 75.

CHAPTER 34 MAXWELL'S EQUATIONS AND ELECTROMAGNETIC WAVES

Section 34-2: Ambiguity in Ampère's Law

Problem

1. A uniform electric field is increasing at the rate of 1.5 V/m·μs. What is the displacement current through an area of 1.0 cm^2 at right angles to the field?

Solution

Maxwell's displacement current is $\varepsilon_0 \partial\phi_E/\partial t = (8.85\times10^{-12}\text{ F/m})(1.5\text{ V/m}\cdot\mu\text{s})(1\text{ cm}^2) = 1.33$ nA. (See Equations 34-1 and 24-2.)

Problem

3. A parallel-plate capacitor of plate area A and spacing d is charging at the rate dV/dt. Show that the displacement current in the capacitor is equal to the conduction current flowing in the wires feeding the capacitor.

Solution

The displacement current is $I_D = \varepsilon_0\partial\phi_E/\partial t$. For a parallel-plate capacitor, $E = q/\varepsilon_0 A$, so $I_D = \varepsilon_0\ \partial(EA)/\partial t = \varepsilon_0\partial(q/\varepsilon_0)/\partial t = dq/dt$. But dq/dt is just the conduction current (the rate at which charge is flowing onto the capacitor plates); hence $I_D = I$.

Problem

5. A parallel-plate capacitor has circular plates with radius 50 cm and spacing 1.0 mm. A uniform electric field between the plates is changing at the rate 1.0 MV/m·s. What is the magnetic field between the plates (a) on the symmetry axis, (b) 15 cm from the axis, and (c) 150 cm from the axis?

Solution

(a) As explained in Example 34-1, cylindrical symmetry and Gauss's law for magnetism require that the **B**-field lines be circles around the symmetry axis, as in Fig. 34-5. For a radius, r, less than the radius of the plates, R, the displacement current is $I_D = \varepsilon_0 d\phi_E/dt = \varepsilon_0(d/dt)\int \mathbf{E}\cdot dA = \varepsilon_0\pi r^2(dE/dt)$, where the integral is over a disk of radius r centered between the plates. Maxwell's form of Ampère's law gives $\oint \mathbf{B}\cdot d\ell = 2\pi rB = \mu_0 I_D$, where the line integral is around the circumference of the disk. Thus, $B = \frac{1}{2}\mu_0\varepsilon_0 r(dE/dt) = r(dE/dt)/2c^2$, where c is the speed of light (Equation 34-16). On the symmetry axis, $r = 0$, so $B = 0$. (b) For $r = 15\text{ cm} < R$, $B = \frac{1}{2}(0.15\text{ m})(10^6\text{ V/m·s})/(3\times10^8\text{ m/s})^2 = 8.33\times10^{-13}$ T. (c) For $r > R$, the displacement current is $I_D = \varepsilon_0\pi R^2(dE/dt)$, so $B = (dE/dt)R^2/2c^2r$. At $r = 150$ cm, $B = (10^6\text{ V/m·s})(50\text{ cm})^2/2(3\times10^8\text{ m/s})^2\times(150\text{ cm}) = 9.26\times10^{-13}$ T.

Section 34-4: Electromagnetic Waves

Problem

7. At a particular point the instantaneous electric field of an electromagnetic wave points in the $+y$ direction, while the magnetic field points in the $-z$ direction. In what direction is the wave propagating?

Solution

For electromagnetic waves in vacuum, the directions of the electric and magnetic fields, and of wave propagation, form a right-handed coordinate system, as shown. (The vector relationship is summarized in Equation 34-20b.) Therefore, the given wave is headed in the $-x$-direction.

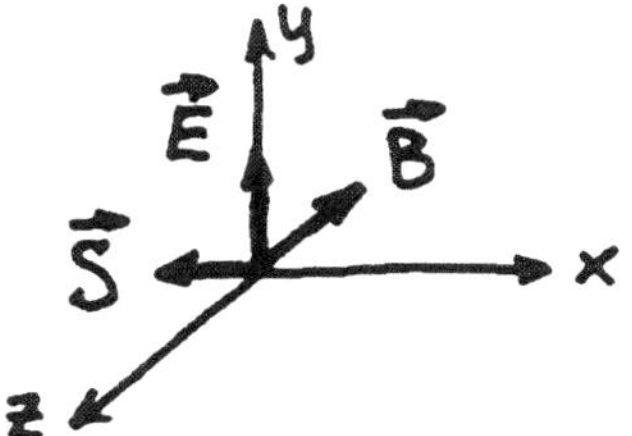

Problem 7 Solution.

Problem

9. The electric field of a radio wave is given by $\mathbf{E} = E\sin(kz-\omega t)(\hat{\imath}+\hat{\jmath})$. (a) What is the peak amplitude of the electric field? (b) Give a unit vector in the direction of the magnetic field at a place and time where $\sin(kz-\omega t)$ is positive.

Solution

(a) The peak amplitude is the magnitude of $E(\hat{\imath}+\hat{\jmath})$, which is $E\sqrt{2}$. Note that $\hat{\imath}+\hat{\jmath}=\sqrt{2}\hat{n}$, where $\hat{\mathbf{n}}$ is a unit vector 45° between the positive x and y axes.
(b) When $\mathbf{E}$ is parallel to $\hat{\mathbf{n}}$ (for $\sin(kz-\omega t)$ positive) $\mathbf{B}$ points 45° into the second quadrant (so that $\mathbf{E}\perp\mathbf{B}$, and $\mathbf{E}\times\mathbf{B}$ is in the $+z$ direction). Thus, $\mathbf{B}$ is parallel to the unit vector $(-\hat{\imath}+\hat{\jmath})/\sqrt{2}$.

Problem

11. Show that it is impossible for an electromagnetic wave in a vacuum to have a time-varying component of its electric field in the direction of its magnetic field. *Hint:* Assume $\mathbf{E}$ does have such a component, and show that you cannot satisfy both Gauss and Faraday.

Solution

Consider a wave propagating in the x direction through a vacuum (no charges or currents present). Gauss's laws for electricity and magnetism require the field lines to continue forever in the y-z plane (no E_x or B_x). We may choose the z direction parallel to the magnetic field, $\mathbf{B}=B_z\hat{\mathbf{k}}$. Suppose $\mathbf{E}=E_y\hat{\jmath}+E_z\hat{\mathbf{k}}$. The discussion of Faraday's law leading to Equation 34-12b shows that $\partial E_y/\partial x=-\partial B_z/\partial t$. But consider a corresponding loop in the x-z plane. Then $-\partial E_z/\partial x=-\partial B_y/\partial t=0$, since there is no B_y by assumption. Because all the space-time dependence in the wave occurs in the combination $kx-\omega t$ (see Problem 10), $\omega\partial E_z/\partial t=-k\partial E_z/\partial x=0$. Then E_z must be a constant and is therefore not a part of the wave. (Similar consideration of Ampère's law over the loop in the x-y plane gives $\varepsilon_0\mu_0\partial E_z/\partial t=0$ directly, with the same conclusion regarding E_z.)

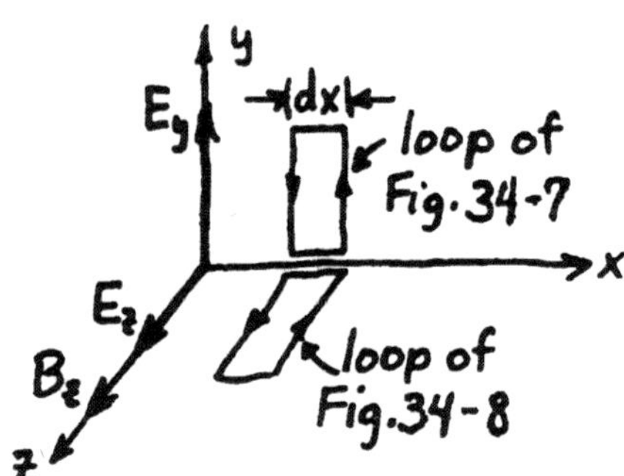

Problem 11 Solution.

Section 34-5: The Speed of Electromagnetic Waves

Problem

13. Your intercontinental telephone call is carried by electromagnetic waves routed via a satellite in geosynchronous orbit at an altitude of 36,000 km. Approximately how long does it take before your voice is heard at the other end?

Solution

Assuming the satellite is approximately overhead, we can estimate the round-trip travel time by $\Delta t=\Delta r/c=(2\times 36{,}000\text{ km})/(3\times 10^5\text{ km/s})=0.24$ s.

Problem

15. Roughly how long does it take light to go 1 foot?

Solution

In $\Delta t=1$ ns, light travels about $\Delta r=c\,\Delta t=(3\times 10^8\text{ m/s})(10^{-9}\text{ s})=30$ cm, or about one foot.

Problem

17. "Ghosts" on a TV screen occur when part of the signal goes directly from transmitter to receiver, while part takes a longer route, reflecting off mountains or buildings (Fig. 34-31). The electron beam in a 50-cm-wide TV tube "paints" the picture by scanning the beam from left to right across the screen in about 10^{-4} s. If a "ghost" image appears displaced about 1 cm from the main image, what is the difference in path lengths of the direct and indirect signals?

Solution

The time it takes for the electron beam to sweep across 1 cm of the TV screen is $1\text{ cm}/(50\text{ cm}/10^{-4}\text{ s})=2\times 10^{-6}$ s, which equals the time delay of the ghost signal. Therefore, the path difference is $\Delta r=c\,\Delta t=(3\times 10^8\text{ m/s})(2\times 10^{-6}\text{ s})=600$ m.

Problem

19. Problem 69 shows that the speed of electromagnetic waves in a transparent dielectric is given by $1/\sqrt{\kappa\varepsilon_0\mu_0}$, where κ is the dielectric constant described in Chapter 26. An experimental measurement gives 1.97×10^8 m/s for the speed of light in a piece of glass. What is the dielectric constant of this glass at optical frequencies?

Solution

Since $c=1/\sqrt{\varepsilon_0\mu_0}$ (speed of light in vacuum), we can write $v=c/\sqrt{\kappa}$ for the speed of light in a dielectric.

Then $\kappa = (c/v)^2 = (3/1.97)^2 = 2.32$. (Note: $\sqrt{\kappa}$ at optical frequencies is called the index of refraction; see Section 35-3.)

Section 34-6: Properties of Electromagnetic Waves

Problem

21. A 60-Hz power line emits electromagnetic radiation. What is the wavelength?

Solution

The wavelength in a vacuum (or air) is $\lambda = c/f = (3\times10^8 \text{ m/s})/(60 \text{ Hz}) = 5\times10^6$ m, almost as large as the radius of the Earth.

Problem

23. A CB radio antenna is a vertical rod 2.75 m high. If this length is one-fourth of the CB wavelength, what is the CB frequency?

Solution

From Equation 34-17b, $f = c/\lambda = (3\times10^8 \text{ m/s})\div(4\times2.75 \text{ m}) = 27.3$ MHz.

Problem

25. What would be the electric field strength in an electromagnetic wave whose magnetic field equaled that of Earth, about 50 μT?

Solution

For a wave in free space, Equation 34-18 gives $E = cB = (3\times10^8 \text{ m/s})(0.5\times10^{-4} \text{ T}) = 15 \text{ kV/m}$.

Problem

27. A radio receiver can detect signals with electric fields as low as 320 μV/m. What is the corresponding magnetic field?

Solution

From Equation 34-18 for waves in free space, $B = E/c = (320 \ \mu\text{V/m})/(3\times10^8 \text{ m/s}) = 1.07$ pT.

Section 34-8: Polarization

Problem

29. Polarized light is incident on a sheet of polarizing material, and only 20% of the light gets through. What is the angle between the electric field and the polarization axis of the material?

Solution

From the law of Malus (Equation 34-19), $S/S_0 = \cos^2\theta = 20\%$, or $\theta = \cos^{-1}(\sqrt{0.2}) = 63.4°$.

Problem

31. A polarizer blocks 75% of a polarized light beam. What is the angle between the beam's polarization and the polarizer's axis?

Solution

Equation 34-19 gives $\theta = \cos^{-1}\sqrt{S/S_0} = \cos^{-1}\sqrt{1-75\%} = \cos^{-1}\sqrt{\frac{1}{4}} = 60°$.

Problem

33. Unpolarized light of intensity S_0 passes first through a polarizer with its polarization axis vertical, then through one with its axis at 35° to the vertical. What is the light intensity after the second polarizer?

Solution

Only 50% (one half the intensity) of the unpolarized light is transmitted through the first polarizer, and the second cuts this down by $\cos^2 35°$. Therefore $\frac{1}{2}\cos^2 35° = 33.6\%$ of the unpolarized intensity gets through both polarizers.

Problem

35. Unpolarized light with intensity S_0 passes through a stack of five polarizing sheets, each with its axis rotated 20° with respect to the previous one. What is the intensity of the light emerging from the stack?

Solution

Only $\frac{1}{2}$ (or 50%) of the incident unpolarized intensity gets through the first polarizing sheet, while $\cos^2 20°$ (or 88.3%) is passed through each of the succeeding four sheets. The net percentage emerging is $\frac{1}{2}(\cos^2 20°)^4 = 30.4\%$.

Problem

37. Polarized light with average intensity S_0 passes through a sheet of polarizing material which is rotating at 10 rev/s. At time $t = 0$ the polarization axis is aligned with the incident polarization. Write an expression for the transmitted intensity as a function of time.

Solution

Because the frequency of light is much greater than that of the rotating polarizer (5×10^{14} Hz $\gg$ 10 Hz), the law of Malus relates the average light intensities (see discussion leading to Equation 34-21a). Thus, $S = S_0\cos^2\theta$. For $\theta = \omega t$, where $\omega = 2\pi\times10 \text{ s}^{-1}$, $S = S_0\cos^2(20\pi \text{ s}^{-1})t = \frac{1}{2}S_0[1 + \cos(40\pi \text{ s}^{-1})t]$.

Section 34-10: Energy in Electromagnetic Waves

Problem

39. What would be the average intensity of a laser beam so strong that its electric field produced dielectric breakdown of air (which requires $E_p = 3\times 10^6$ V/m)?

Solution

Equation 34-21b for the average intensity of electromagnetic waves gives $\bar{S} = E_p^2/2\mu_0 c =$ $(3\times 10^6 \text{ V/m})^2(8\pi\times 10^{-7} \text{ H/m})^{-1}(3\times 10^8 \text{ m/s})^{-1} =$ 11.9 GW/m^2.

Problem

41. A radio receiver can pick up signals with peak electric fields as low as 450 μV/m. What is the average intensity of such a signal?

Solution

From Equation 34-21b, $\bar{S} = E_p^2/2\mu_0 c =$ $(450\ \mu\text{V/m})^2/(8\pi\times 10^{-7} \text{ H/m})(3\times 10^8 \text{ m/s}) =$ $2.69\times 10^{-10} \text{ W/m}^2$.

Problem

43. A laser blackboard pointer delivers 0.10 mW average power in a beam 0.90 mm in diameter. Find (a) the average intensity, (b) the peak electric field, and (c) the peak magnetic field.

Solution

(a) If the average power is spread uniformly over the beam area, $\bar{S} = 0.1 \text{ mW}/\frac{1}{4}\pi(0.9 \text{ mm})^2 = 157 \text{ W/m}^2$. (b) Equation 34-21b gives $E_p = \sqrt{2\mu_0 c\bar{S}} = 344$ V/m, and (c) Equation 34-18 gives $B_p = E_p/c = 1.15\ \mu$T.

Problem

45. The United States' safety standard for continuous exposure to microwave radiation is 10 mW/cm^2. The glass door of a microwave oven measures 40 cm by 17 cm and is covered with a metal screen that blocks microwaves. What fraction of the oven's 625-W microwave power can leak through the door window without exceeding the safe exposure to someone right outside the door? Assume the power leaks uniformly through the window area.

Solution

The power corresponding to the safety standard of intensity, uniformly distributed over the window area, is $(10 \text{ mW/cm}^2)(40\times 17 \text{ cm}^2) = 6.8$ W, which is 1.09% of the microwave's 625 W power output.

Problem

47. Use the fact that sunlight intensity at Earth's orbit is 1368 W/m^2 to calculate the Sun's total power output.

Solution

If the Sun is emitting isotropically, its power output is $\mathcal{P} = 4\pi r^2\bar{S}$ (from Equation 34-22). Using values of comparable accuracy for the Earth's average orbital distance, we find $\mathcal{P} = 4\pi(1.496\times 10^{11} \text{ m})^2\times$ $(1368 \text{ W/m}^2) = 3.85\times 10^{26}$ W.

Problem

49. During its 1989 encounter with Neptune, the Voyager II spacecraft was 4.5×10^9 km from Earth. Its images of Neptune were broadcast by a radio transmitter with a mere 21-W average power output. What would be (a) the average intensity and (b) the peak electric field received at Earth if the transmitter broadcast equally in all directions? (The received signal was actually somewhat stronger because Voyager used a directional antenna.)

Solution

(a) The average intensity at a distance r from an isotropic emitter is (Equation 34-22) $\bar{S} = \mathcal{P}/4\pi r^2 =$ $21 \text{ W}/4\pi(4.5\times 10^{12} \text{ m})^2 = 8.25\times 10^{-26} \text{ W/m}^2$. (b) This corresponds to a peak electric field of only (Equation 34-21b) $E_p = \sqrt{2\mu_0 c\bar{S}} = 7.89\times 10^{-12}$ V/m. (The one-way travel time queried in the caption of Fig. 34-32 is $r/c = 4$ h, 10 min.)

Problem

51. At 1.5 km from the transmitter, the peak electric field of a radio wave is 350 mV/m. (a) What is the transmitter's power output, assuming it broadcasts uniformly in all directions? (b) What is the peak electric field 10 km from the transmitter?

Solution

(a) Equations 34-22 and 21b can be combined to express the average power output of an isotropic transmitter in terms of the peak electric field at a distance r, $\mathcal{P} = 4\pi r^2(E_p^2/2\mu_0 c) = (1.5 \text{ km})^2\times$ $(350 \text{ mV/m})^2/(2\times 10^{-7}\times 3\times 10^8 \text{ H/s}) = 4.59$ kW. (b) Since $r^2E_p^2$ is a constant, $E_p' = (r/r')E_p =$ $(1.5/10)(350 \text{ mV/m}) = 52.5$ mV/m at a distance of 10 km.

Problem

53. A typical fluorescent lamp is a little over 1 m long and a few cm in diameter. How do you expect the light intensity to vary with distance (a) near the lamp but not near either end and (b) far from the lamp?

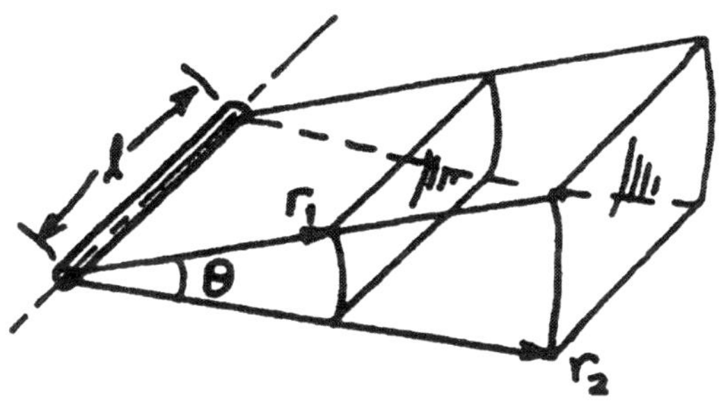

Problem 53 Solution.

Solution

(a) Near the lamp, but far from its ends, light waves travel approximately radially outwards from the tube axis. The power crossing two co-axial cylindrical patches is the same, but the area of each patch is proportional to the radius. Therefore, the intensity varies as $1/r$. ($S_1A_1 = S_2A_2 = S_1\theta r_1\ell = S_2\theta r_2\ell$.) This is the same relation as depicted in the solution to Problem 16-44. (b) Very far away, the lamp appears as a point source, and Equation 34-22 holds, so the intensity varies like $1/r^2$.

Section 34-11: Wave Momentum and Radiation Pressure

Problem

55. What is the radiation pressure exerted on a light-absorbing surface by a laser beam whose intensity is 180 W/cm^2?

Solution

The radiation pressure generated by a totally absorbed electromagnetic wave of given average intensity (from Equation 34-24) is $P_{\text{rad}} = \bar{S}/c = (180\ \text{W/cm}^2)/(3\times10^8\ \text{m/s}) = 6$ mPa.

Problem

57. The average intensity of noonday sunlight is about 1 kW/m^2. What is the radiation force on a solar collector measuring 60 cm by 2.5 m if it is oriented at right angles to the incident light and absorbs all the light?

Solution

For sunlight incident normally on a perfect absorber, $P_{\text{rad}} = \bar{S}/c$ (Equation 34-24). Therefore, the force on the solar collector is $P_{\text{rad}}A = (1\ \text{kW/m}^2)\times(0.6\times2.5\ \text{m}^2)/(3\times10^8\ \text{m/s}) = 5\ \mu\text{N}$.

Problem

59. A 65-kg astronaut is floating in empty space. If the astronaut shines a 1.0-W flashlight in a fixed direction, how long will it take the astronaut to accelerate to a speed of 10 m/s?

Solution

The reaction force of the light emitted on the flashlight equals the rate at which momentum is carried away by the beam, or $F = dp/dt = (dU/dt)/c = \mathcal{P}/c$. Such a force could accelerate a mass m from rest to a speed v in time $t = v/a = mv/F = mcv/\mathcal{P}$. For the values given for the astronaut and flashlight, $t = (65\ \text{kg})\times(3\times10^8\ \text{m/s})(10\ \text{m/s})/(1\ \text{W}) = 1.95\times10^{11}\ \text{s} = 6.18\times10^3$ y (impractically long).

Paired Problems

Problem

61. Find the peak electric and magnetic fields 1.5 m from a 60-W light bulb that radiates equally in all directions.

Solution

For an isotropic source of electromagnetic waves (in a medium with vacuum permittivity and permeability) Equations 34-21b and 22 give $\bar{S} = \mathcal{P}/4\pi r^2 = E_p^2/2\mu_0c$, therefore $E_p = \sqrt{2\mu_0c\mathcal{P}/4\pi r^2} = (2\times10^{-7}\times 3\times10^8\times60)^{1/2}(\text{ V/m})/(1.5\ \text{m}) = 40$ V/m. Then Equation 34-18 gives $B_p = E_p/c = 133$ nT.

Problem

63. Unpolarized light is incident on two polarizers with their axes at 45°. What fraction of the incident light gets through?

Solution

50% of the incident unpolarized light (a random mixture of all polarizations) is transmitted by the first polarizer, and $\cos^2 45° = 50\%$ of that (see Equation 34-19) by the second, for a total of $0.5\times0.5 = 25\%$ for both.

Problem

65. What is the radiation force on the door of a microwave oven if 625 W of microwave power hits the door at right angles and is reflected?

Solution

The radiation pressure for normally incident, perfectly reflected electromagnetic waves is $2\bar{S}/c$. We suppose that the microwave power is uniformly spread over the area of the door (as for plane waves), so $\bar{S} = \mathcal{P}/A$. Then the force on the door is simply $\mathcal{P}_{\text{rad}}A = (2\bar{S}/c)A = 2\mathcal{P}/c = 2(625\text{ W})/(3\times10^8\text{ m/s}) = 4.17\ \mu\text{N}$.

Problem

67. A 60-W light bulb is 6.0 cm in diameter. What is the radiation pressure on an opaque object at the bulb's surface?

Solution

Suppose that the filament in the bulb is small enough to be considered as an isotropic point source. Then the radiation pressure on an opaque (perfectly absorbing) object is $P_{\text{rad}} = \bar{S}/c$ and $\bar{S} = \mathcal{P}/4\pi r^2$. Thus $P_{\text{rad}} = (60\text{ W}/4\pi)(3\text{ cm})^{-2}(3\times10^8\text{ m/s})^{-1} = 17.7\ \mu\text{Pa}$.

Supplementary Problems

Problem

69. Maxwell's equations in a dielectric resemble those in vacuum (Equations 34-6 through 34-9), but with ϕ_E in Ampère's law replaced by $\kappa\phi_E$, where κ is the dielectric constant introduced in Chapter 26. Show that the speed of electromagnetic waves in such a dielectric is $c/\sqrt{\kappa}$.

Solution

The effect of a linear, isotropic, homogeneous dielectric medium in Gauss's law is to replace ε_0 by $\kappa\varepsilon_0$. Maxwell defined the displacement current in a dielectric analogously, as $\kappa\varepsilon_0 d\phi_E/dt$. Therefore, Maxwell's equations in a dielectric medium (containing no free charge or conduction currents) are just those in Table 34-2 with $\kappa\varepsilon_0$ replacing ε_0. The discussion in Sections 34-4 through 6 applies to waves in a dielectric medium, with the same replacement. In particular, the wave speed (Equation 34-16) becomes $1/\sqrt{\kappa\varepsilon_0\mu_0} = c/\sqrt{\kappa}$. (In Section 35-3, $\sqrt{\kappa}$ is defined as the index of refraction of the medium.)

Problem

71. A radar system produces pulses consisting of 100 full cycles of a sinusoidal 70-GHz electromagnetic wave. The average power while the transmitter is on is 45 MW, and the waves are confined to a beam 20 cm in diameter. Find (a) the peak electric field, (b) the wavelength, (c) the total energy in a pulse, and (d) the total momentum in a pulse. (e) If the transmitter produces 1000 pulses per second, what is its average power output?

Solution

(a) The average intensity of a pulse is the average power during a pulse divided by the beam area, $\bar{S} = \mathcal{P}/\pi R^2$, and the peak electric field is therefore $E_p = \sqrt{2\mu_0 c\bar{S}} = (1/R)\sqrt{2\mu_0 c\mathcal{P}/\pi} = [(8\times10^{-7}\text{ H/m})\times(3\times10^8\text{ m/s})(45\text{ MW})]^{1/2}/(0.1\text{ m}) = 1.04\text{ MV/m}$. (b) The wavelength is $\lambda = c/f = (3\times10^8\text{ m/s})\div(70\text{ GHz}) = 4.29\text{ mm}$. (c) 100 full cycles has a duration of $100/f = 100/(70\text{ GHz}) = 1.43\text{ ns}$, so the total energy in a pulse is $\mathcal{P}t = (45\text{ MW})(1.43\text{ ns}) = 64.3\text{ mJ}$. (d) From Equation 34-23, the momentum per pulse is $(64.3\text{ mJ})/(3\times10^8\text{ m/s}) = 2.14\times10^{-10}\text{ kg}\cdot\text{m/s}$. (e) The average power output (which includes the time between pulses) is $(1000\text{ pulses/s})(64.3\text{ mJ/pulse}) = 64.3\text{ W}$. (This is, of course, much less than the average power during a pulse, since pulses are emitted for only $10^3\times1.43\text{ ns} = 1.43\ \mu\text{s}$ each second. Thus $\mathcal{P}_{\text{out}} = \mathcal{P}_{\text{pulse}}\times1.43\ \mu\text{s/s}$, as above.)

Problem

73. The peak electric field measured at 8.0 cm from a light source is 150 W/m^2, while at 12 cm it measures 122 W/m^2. Describe the shape of the source.

Solution

The intensity is proportional to the square of the peak electric field (Equation 34-21b), so the given data implies that the ratio of intensities is proportional to that of the inverse distances, $(150/122)^2 = 1.51 \approx (12/8)$. Thus, $\bar{S}r$ is roughly constant. The intensity near a long, cylindrically symmetric source, where r is the axial distance, has this space dependence (see Problem 53).

Problem

75. Studies of the origin of the solar system suggest that sufficiently small particles might be blown out of the solar system by the force of sunlight. To see how small such particles must be, compare the force of sunlight with the force of gravity, and solve for the particle radius at which the two are equal. Assume the particles are spherical and have density 2 g/cm^3. Why do you not need to worry about the distance from the Sun.

Solution

The force due to radiation pressure (away from the Sun) on a totally absorbing spherical particle (radius R, cross-sectional area πR^2) is $P_{\text{rad}}\pi R^2$, where

$P_{rad} = \bar{S}/c$, and $\bar{S} = \mathcal{P}_\odot/4\pi r^2$ is the intensity of Sunlight at a distance r from the Sun. The force of the Sun's gravity on the particle (toward the Sun) is $GM_\odot(\frac{4}{3}\pi R^3\rho)/r^2$, where ρ is the particle's density. The two forces are equal in magnitude for particles with

$$R\rho = 3\mathcal{P}_\odot/16\pi \mathrm{G}M_\odot c = \frac{3(3.85\times10^{26}\ \mathrm{W}/16\pi)}{(6.67\times10^{-11}\ \mathrm{N{\cdot}m^2/kg^2})(1.99\times10^{30}\ \mathrm{kg})(3\times10^{8}\ \mathrm{m/s})} = 5.77\times10^{-5}\ \mathrm{g/cm^2}.$$

(Note that since both forces are proportional to r^{-2}, the result is independent of r.) For particles with density 2 g/cm^3, $R = 2.89\times10^{-5}$ cm $= 0.289\ \mu$m. Particles of the same density but smaller radius would be swept away by the solar radiation pressure.

PART 4 CUMULATIVE PROBLEMS

Problem

1. An air-insulated parallel-plate capacitor has plate area 100 cm^2 and spacing 0.50 cm. The capacitor is charged and then disconnected from the charging battery. A thin-walled, nonconducting box of the same dimensions as the capacitor is filled with water at 20.00°C. The box is released at the edge of the capacitor and moves without friction into the capacitor (Fig. 1). When it reaches equilibrium the water temperature is 21.50°C. What was the original voltage on the capacitor?

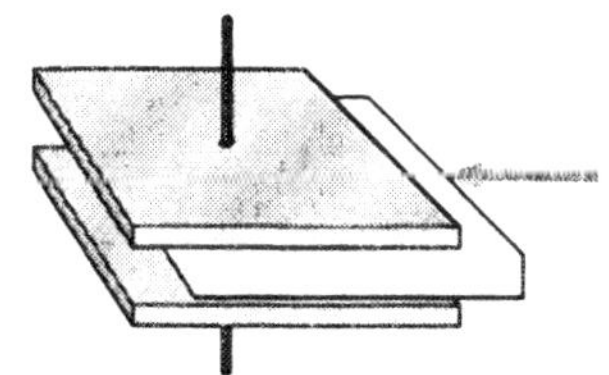

FIGURE 1 Cumulative Problem 1.

Solution

We assume that the entire difference between the initial and final values of the electrostatic energy stored in the capacitor is eventually dissipated as heat in the water (the dielectric medium), as mentioned following Equation 26-12. Thus, $U_0 - U = mc\,\Delta T$. From Equation 26-12, $U_0 - U = U_0 - U_0/\kappa = \frac{1}{2}C_0V_0^2(\kappa-1)/\kappa$, where $C_0 = \varepsilon_0 A/d = (8.85\ \mathrm{pF/m})\times(10^{-2}\ \mathrm{m^2})/(5\times10^{-3}\ \mathrm{m}) = 17.7$ pF, and $\kappa = 78$ for water; see Table 26-1. The amount of water is $m = (1\ \mathrm{g/cm^3})(100\ \mathrm{cm^2})(0.5\ \mathrm{cm}) = 50$ g, so putting everything together and solving for V_0, we find:

$$V_0 = \sqrt{\frac{2mc\,\Delta T}{C_0(\kappa-1)/\kappa}} = \sqrt{\frac{2(50\ \mathrm{g})(4.184\ \mathrm{J/g{\cdot}°C})(1.5°\mathrm{C})}{(17.7\ \mathrm{pF})(77/78)}} = 5.99\ \mathrm{MV}.$$

Problem

3. Five wires of equal length 25 cm and resistance 10 Ω are connected, as shown in Fig. 3. Two solenoids, each 10 cm in diameter, extend a long way perpendicular to the page. The magnetic fields of both solenoids point out of the page; the field strength in the left-hand solenoid is increasing at 50 T/s while that in the right-hand solenoid is decreasing at 30 T/s. Find the current in the resistance wire shared by both triangles. Which way does the current flow?

Solution

The increasing (decreasing) magnetic flux in the left-hand (right-hand) triangle in Fig. 3 gives rise to a clockwise (counterclockwise) induced emf in loop 1 (2), so the circuit behaves like the one shown in Fig. 28-18, but with both batteries reversed. Then the solution to

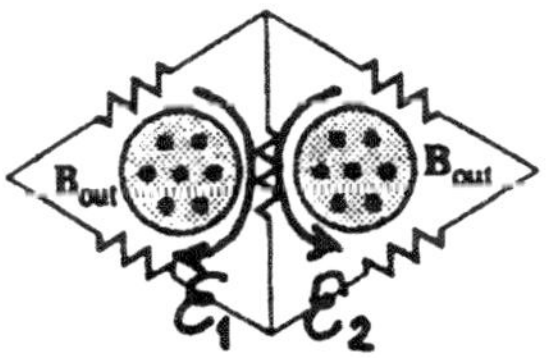

FIGURE 3 Cumulative Problem 3 Solution.

Problem 28-33 gives the current in the middle wire as

$$I_3 = -(\mathcal{E}_2 R_1 + \mathcal{E}_1 R_2)/(R_1 R_2 + R_2 R_3 + R_3 R_1).$$

The minus sign means that the direction of the current is opposite to that shown in Fig. 28-18, i.e., downward. There are two wires in the left and right branches and one in the middle, so the corresponding resistances are $R_1 = R_2 = 2R_3 = 2(10\ \Omega)$. Then I_3 becomes $-(\mathcal{E}_1 + \mathcal{E}_2)/4(10\ \Omega)$. The magnitudes of the emf's are given by Faraday's law, $\mathcal{E} = |d\phi/dt| = \pi r^2\,|dB/dt|$, and both solenoids have the same cross-sectional area, so $I_3 = -\pi(5\text{ cm})^2(50\text{ T/s} + 30\text{ T/s})/4(10\ \Omega) = -15.7\text{ mA}$.

Problem

5. A coaxial cable consists of an inner conductor of radius a and an outer conductor of radius b; the space between the conductors is filled with insulation of dielectric constant κ (Fig. 4). The cable's axis is the z axis. The cable is used to carry electromagnetic energy from a radio transmitter to a broadcasting antenna. The electric field between the conductors points radially from the axis, and is given by $E = E_0(a/r)\cos(kz - \omega t)$. The magnetic field encircles the axis, and is given by $B = B_0(a/r)\cos(kz - \omega t)$. Here E_0, B_0, k, and ω are constants. (a) Show, using appropriate closed surfaces and loops, that these fields satisfy Maxwell's equations. Your result shows that the cable acts as a "waveguide," confining an electromagnetic wave to the space between the conductors. (b) Find an expression for the speed at which the wave propagates along the cable.

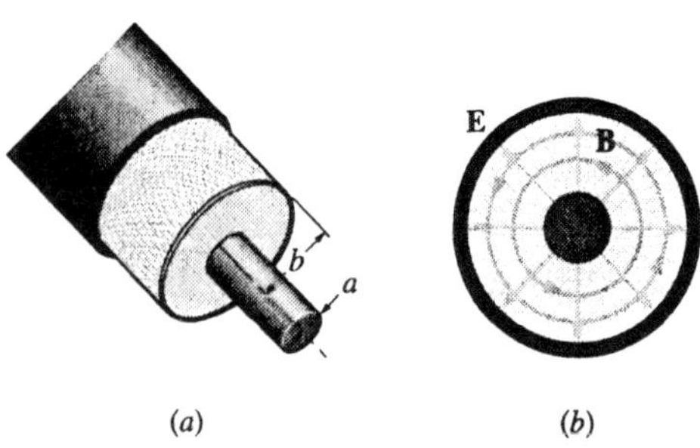

FIGURE 4 Cumulative Problem 5.

Solution

The method for (a) demonstrating that the given fields in a coaxial cable (so-called transverse electric and magnetic waves, or TEM-waves) satisfy Maxwell's equations and for (b) finding their speed is the same as that used in Sections 34-4 and 5, except that cylindrical coordinates r, θ, z, and Maxwell's equations in the dielectric (see solution to Problem 34-69) must be used. Consider an infinitesimal volume of dielectric at a point (r, θ, z) bounded by mutually orthogonal coordinate displacements $\mathbf{dr}, r\ \mathbf{d\theta}, \mathbf{dz}$. The dielectric contains no free charges or conduction currents, so the given radial electric field and circulating magnetic field satisfy Gauss's laws for electricity and magnetism in the dielectric.

In Faraday's law, take a CCW loop and sides dr and dz, at fixed θ, so that the normal to the area it bounds is parallel to $\mathbf{B}$, as shown. Then, since $\mathbf{E} \perp \mathbf{dz}$ and $\mathbf{B} \perp \mathbf{dr}$ and $\mathbf{dz}$,

$$\oint \mathbf{E}\cdot d\ell = -E\ dr + \left(E + \frac{\partial E}{\partial z}\ dz\right) dr = \frac{\partial E}{\partial z}\ dz\ dr$$
$$= -\frac{d\phi_B}{dt} = -\frac{d}{dt}(B\ dz\ dr) = -\frac{\partial B}{\partial t}\ dz\ dr,$$

or $\partial E/\partial z = -(E_0 ak/r)\sin(kz - \omega t) = -\partial B/\partial t = -(B_0 a\omega/r)\sin(kz - \omega t)$. Thus, $E_0 k = B_0\omega$ (same as Equation 34-14).

In Ampere's law, take a CCW loop with sides $r\ d\theta$ and dz, at fixed r, so that the normal to its area is parallel to $\mathbf{E}$. Then, since $\mathbf{B} \perp \mathbf{dz}$ and $\mathbf{E} \perp r\ \mathbf{d\theta}$ and $\mathbf{dz}$,

$$\oint \mathbf{B}\cdot d\ell = Br\ d\theta - \left(B + \frac{\partial B}{\partial z}\ dz\right) r\ d\theta = -\frac{\partial B}{\partial z} r\ d\theta\ dz$$
$$= \kappa\varepsilon_0\mu_0\frac{d\phi_E}{dt} = \kappa\varepsilon_0\mu_0\frac{d}{dt}(Er\ d\theta\ dz)$$
$$= \kappa\varepsilon_0\mu_0\frac{\partial E}{\partial t} r\ d\theta\ dz,$$

or $-\partial B/\partial z = (B_0 ak/r)\sin(kz - \omega t) = \kappa\varepsilon_0\mu_0 \times (\partial E/dt) = \kappa\varepsilon_0\mu_0(E_0 a\omega/r)\sin(kz - \omega t)$. Thus, $B_0 k = \kappa\varepsilon_0\mu_0 E_0\omega$ (same as Equation 34-15 except for κ).

Therefore, Maxwell's equations are satisfied provided E_0, B_0, k and ω are related by $B_0\omega = E_0 k$ and $B_0 k = E_0\omega\kappa\varepsilon_0\mu_0$. The wave speed follows by dividing these equations to eliminate the amplitudes. Then $\omega/k = k/\omega\kappa\varepsilon_0\mu_0$, or $(\omega/k)^2 = 1/\kappa\varepsilon_0\mu_0 = v^2$, and $v = 1/\sqrt{\kappa\varepsilon_0\mu_0} = c/\sqrt{\kappa}$.

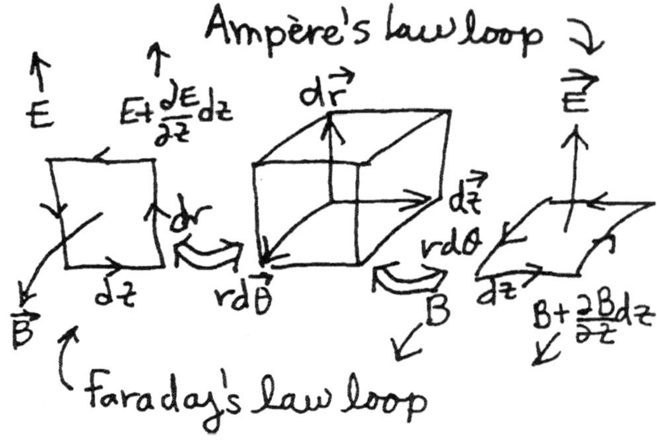

Cumulative Problem 5 Solution.